VIOLENT NON-STATE ACTORS IN MODERN CONFLICT

SANDHURST TRENDS IN INTERNATIONAL CONFLICT

Edited by David Brown, Donette Murray, Malte Riemann, Norma Rossi and Martin A. Smith, The Royal Military Academy Sandhurst, UK

The Sandhurst Trends in International Conflict series is a cutting-edge forum and platform for original thought and debate on military and security matters within the contemporary international security environment. It aims to stimulate authors to think critically about contemporary conflict and security more generally and to identify and evaluate practical, political and doctrinal lessons from recent experience. The Sandhurst series invites practitioners and academics from the fields of security, diplomacy, the law, politics and the military to interrogate and publish on the key debates that will shape both the contemporary international security environment and a modern military operating within it.

Other books in this series

War Amongst the People: Critical Assessments
978 1 912440 02 3

Fragile and Failing States: Challenges and Responses
978 1 912440 19 1

Violent Non-State Actors in Modern Conflict

Edited by
DAVID BROWN
DONETTE MURRAY
MALTE RIEMANN
NORMA ROSSI
MARTIN A. SMITH

Howgate Publishing Limited

First published in 2021 by
Howgate Publishing Limited
Station House
50 North Street
Havant
Hampshire
PO9 1QU
Email: info@howgatepublishing.com
Web: www.howgatepublishing.com

British Library Cataloguing-in-Publication Data
A catalogue record for this book is available from the British Library

ISBN 978-1-912440-20-7 (pbk)
ISBN 978-1-912440-21-4 (ebk - PDF)
ISBN 978-1-912440-22-1 (ebk - ePUB)

The views expressed in this book are those of the individual authors and do not necessarily reflect official policy or position.

CONTENTS

FIGURE

ABOUT THE EDITORS

Dr David Brown is a Senior Lecturer in the Department of Defence and International Affairs at the Royal Military Academy Sandhurst. He has written extensively on a range of security-related issues, publishing books and articles on US and UK foreign and defence policy, contemporary power relations, aspects of European security and international intervention.

Dr Donette Murray is a Senior Lecturer at the Royal Military Academy Sandhurst. A Fellow of the Supreme Headquarters Allied Powers Europe (SHAPE) and a former political advisor, she holds a doctorate from the University of Ulster and an LLM in International Law from the University of Maastricht. The author of four books on US foreign policy, she has also published in the *Hague Yearbook of International Law* on states' use of force in self-defence against non-state actors.

Dr Malte Riemann is a Senior Lecturer in the Department of Defence and International Affairs at the Royal Military Academy Sandhurst. He studied in Bremen and Pietermaritzburg and holds a PhD in International Relations from the University of Reading. His fields of interest include the privatisation of war, urban violence, the medicalisation of security, and the historicity of non-state actors.

Dr Norma Rossi is a Senior Lecturer in Defence and International Affairs at the Royal Military Academy Sandhurst. She studied in Rome and Paris and received her PhD in Politics and International Relations from the University of Reading on an Earhart Foundation fellowship. Her research interests include transnational organised crime, the role of far-right parties in security discourses, the changing character of conflict, and the role of Professional Military Education in Security Sector Reform.

Dr Martin A. Smith is Senior Lecturer in Defence and International Affairs at the Royal Military Academy Sandhurst. Prior to joining RMAS he was at the Department of Peace Studies University of Bradford, from where he received his PhD in 1994. His main research interests are in the fields of international power, European security and US foreign policy and he has published widely in these areas.

CONTRIBUTORS

Julien Bastrup-Birk serves as Head of Transnational Threats in the UK's Stabilisation Unit, having previously worked in a range of analytical, policy and advisory-related roles. As such, he deploys frequently to fragile and conflicted states around the world to provide advice on interventions and responses to complex cross-border challenges. He holds a Masters in War Studies from King's College, London and is a Doctoral candidate in International Security at the University of Leiden.

Lt Col Grant Davies was previously a reservist Infantry Officer and has served as a British Army Legal Services officer for over 20 years. He has served on operations in Northern Ireland, Kosovo and Afghanistan. He has fulfilled a number of roles, including the senior legal advisor in Regional Command (South) and team leader of the Detention Oversight Team (both in Afghanistan). A graduate in International Relations and Strategic Studies from UCW, Aberystwyth, he also holds a master's degree in Public International Law (King's College London). He is a solicitor advocate and holds a judicial appointment as an assistant Coroner in the jurisdictions Dorset and Wiltshire.

Dr James Fargher specialises in the military history and politics of the Red Sea region. He has worked as an analyst for JANE's as well as security consultancies in London and Singapore. James holds a PhD in War Studies from King's College London.

Dr Helle Malmvig is Senior Researcher at The Danish Institute for International Studies (DIIS), She has published articles and books on Middle East politics, state sovereignty and intervention, US and EU Middle

East policies, and on conflict and creative resistance in Syria and Lebanon. Helle Malmvig is advising The Danish Ministry of Defence, the Ministry of Foreign Affairs, the Danish Parliament and the EU Commission. Malmvig is a frequently used commentator and speaker on Middle East issues in national and international media (e.g. BBC World, al Jazeera, France3, Le Figaro, Huffington Post, Christian Science Monitor), she is also the Danish EuroMeSco representative. She was co-leader of the EU Horizon2020 research project on Middle East Regional Order MENARA from 2016-2019.

Dr Marina Miron holds a PhD in Strategic Studies in 2019 from the University of New South Wales at the Australian Defence Force Academy, Canberra. She is currently a post-doctoral researcher at the Centre for Military Ethics at King's College London and a visiting researcher in military science at the Centre for Military Education, Bogota, Colombia. Marina's areas of study include strategic theory, theory of war, Clausewitzian thought, current trends in military operations and technology (A2/AD, cyber, AI) and military ethics.

Dr Donette Murray is a Senior Lecturer at the Royal Military Academy Sandhurst. A Fellow of the Supreme Headquarters Allied Powers Europe (SHAPE) and a former political advisor, she holds a doctorate from the University of Ulster and an LLM in International Law from the University of Maastricht. The author of four books on US foreign policy, she has also published in the *Hague Yearbook of International Law* on states' use of force in self-defence against non-state actors.

Helene Olsen is a doctoral candidate in the Department of War Studies at King's College London. She teaches on various courses in the Department of War Studies at King's and at the Joint Services Command and Staff College in Shrivenham. Her research focuses on the privatisation of security and military functions, as well as on the relationship between mercenaries and legitimacy. More specifically, Helene explores the tensions between the extensive use of mercenaries in warfare and the apparent moral opposition to their presence on and off the battlefield. Her work will feature in the forthcoming book *Remote Warfare: Interdisciplinary Perspectives* (2020: E-IR).

Dr Norma Rossi is a Senior Lecturer in Defence and International Affairs at the Royal Military Academy Sandhurst. She studied in Rome and Paris and received her PhD in Politics and International Relations from the

University of Reading on an Earhart Foundation fellowship. Her research interests include transnational organised crime, the role of far-right parties in security discourses, the changing character of conflict, and the role of Professional Military Education in Security Sector Reform.

Dr Malte Riemann is a Senior Lecturer in the Department of Defence and International Affairs at the Royal Military Academy Sandhurst. He studied in Bremen and Pietermaritzburg and holds a PhD in International Relations from the University of Reading. His fields of interest include the privatisation of war, urban violence, the medicalisation of security, and the historicity of non-state actors.

Dr Caroline Varin is Senior Lecturer in International Relations at Regent's University London and CEO of Professors Without Borders. She is a member of the Geneva Consultation for Security Policy and has previous experience working in Intelligence Analysis, the Swiss armed forces and the UNHCR. She has published books on Security in Nigeria, Violent Non-State Actors in Africa, Boko Haram and Mercenaries.

Lt Col Matthew Walker has served over 20 years in the British Army. A specialist in stability and counterinsurgency operations, career highlights have included serving in the US land component headquarters during the Battle for Mosul, as an operational mentor to the Afghan National Army and service in Northern Ireland and the Balkans. While at the British Army's Land Warfare Centre he led the team delivering the army's doctrine for the military contribution to stabilisation, counterinsurgency and humanitarian assistance. He was educated at Imperial College, London, The Royal Military Academy Sandhurst and the Defence Academy of the United Kingdom.

Dr Jack Watling is Research Fellow for Land Warfare at the Royal United Services Institute. His work focusses on future concepts of operation and adversary analysis. He has worked in Iraq, Mali, Rwanda, Brunei and further afield. Jack's PhD examined the evolution of British policy responses to civil war in the early twentieth century.

Abigail Watson is a Conflict and Security Policy Coordinator at Saferworld and was formerly research manager at the Oxford Research Group's Remote Warfare Programme. She researches, writes and presents on the military, legal, and political implications of light-footprint military intervention. This includes seven hard-hitting reports, and over a hundred blogs and briefings on the topic. Her work has been featured in the Times, Strategy Bridge, Just Security, the History Channel and UN Security Council reports.

FOREWORD

It gives me great pleasure to introduce this third edition in the Sandhurst Trends in International Conflict series. As with the previous two books the series editors from the Department of Defence and International Affairs at the Royal Military Academy Sandhurst have brought together contributions from a variety of distinguished thinkers and experts from academia, the military, law and government to help us better understand Violent Non-State Actors (VNSAs) and stimulate our thinking on how we might respond to the complex challenges they present.

This book is timely; the first two decades of this century have seen the rise of threats and interventions by organisations and actors which Western Democracies have struggled to properly understand, categorise and address against previously accepted norms. The British Military and its allies have been intimately involved in deployments across the globe that have sought to counter the threat and malign influence posed by these VNSA's. Indeed, for many of my generation who have served extensively and repeatedly in Iraq, Afghanistan and various states in Africa, VNSAs are the defining characteristic of their professional experience of conflict and conflict prevention. The difficulties and complexities of countering VNSAs that possess particular advantages such as agility, access to the people, knowledge of the ground and, increasingly, technology are all too familiar.

Effective counterterrorism and counter-insurgency strategies have always relied on a clear understanding of the nature, ambitions and methods of the enemy, this book again challenges our thinking in this area. The case studies used to argue that the differences between terrorists, warlords and rebels have progressively disappeared as they merge with one another resonate. The possession or aspiration to control territory, a criterion that previously distinguished rebels and warlords from terrorists is far less clear cut in the era of Daesh and similar VNSAs that overlap in terms of their ambitions and in their ability to challenge the sovereign state. Equally the discussion of legitimacy, the laws and the treatment of combatants previously described as mercenaries along with an account of Russia's specific use of 'proxy' VNSAs is fascinating.

As in any military analysis we must constantly force ourselves to ask the question 'so what?' What does this analysis and discussion of VNSAs mean for our strategy, approach, doctrine, structures and our tactics? How do we exploit our diplomatic, economic, cultural and military levers of national power against a VNSA and how do we transform our capabilities, traditionally rooted predominately in the physical dimension to support rapidly evolving information operations in the future to address this evolving threat?

Sadly, I offer no answers, but I do commend this book to you!

Major General Duncan Capps CBE

ABBREVIATIONS

ACD	Abbas Combat Division
ADF	Allied Democratic Forces
AQIM	Al-Qaeda in the Islamic Maghreb
AU	African Union
CAR	Central African Republic
DRC	Democratic Republic of Congo
EPRDF	Ethiopian People's Revolutionary Democratic Front
EU	European Union
FSA	Free Syrian Army
GNA	Government of National Accord
IAC	international armed conflict
ICC	International Criminal Court
ICRC	International Committee of the Red Cross
IHL	International Humanitarian Law
IS	Islamic State
ISIS	Islamic State in Iraq and Syria
LAS	League of Arab States
LDF	Local Defence Forces
LNA	Libyan National Army
MEJA	Military Extraterritorial Jurisdiction Act
MENA	Middle East and North Africa
NIAC	non-international armed conflict
OAU	Organization of African Unity
OIR	Operation Inherent Resolve
PKK	Kurdistan's Workers Party or PKK (*Partiya Karkerên Kurdistanê*)
PMF	Popular Mobilisation Forces

PMSC	Private Military and Security Company
PSC	Private Security Company
PYD	Democratic Union Party (*Partiya Yekitiya Democrat*)
SDF	Syrian Democratic Forces
STC	Southern Transitional Council
UAE	United Arab Emirates
UN	United Nations
UNSC	UN Security Council
VNSA	Violent Non-State Actor
YPG	People's Protection Units (*Yekîneyên Parastina Gel*)

INTRODUCTION

Dramatis Personae

Donette Murray[1]

> During the last decades of the twentieth century, a new type of organized violence developed [where] the distinction between what is private and what is public, state and non-state, formal and informal, what is done for economic and what for political motives, cannot easily be applied.[2]
>
> Mary Kaldor, *New and Old Wars*

The Elephants in the Room

In January 2020, when the Governments of 12 states, including the P5 (France, Russia, China the United States and the United Kingdom) met in Berlin with High Representatives of the United Nations (UN), the African Union (AU), the European Union (EU) and the League of Arab States (LAS), there was an obvious elephant in the room. Conspicuously absent from the German-led initiative to reboot the anaemic Libyan peace process was the warlord General Khalifa Haftar, the most significant opposition figure to emerge in the wake of Muammar Gaddafi's ouster in 2011. Missing too were a plethora of other organised armed groups ranging from the Wagner Group, a Russian Private Security Company (PSC), to myriad militias (jihadist and non-jihadist; pro-government and anti-government; terrorist and non-proscribed) such as Libya Dawn, Libya Shield, the Zintan, al-Sawaiq, al-Qaqa Brigades, the Libyan Petroleum Facilities Guard, the 17 February Brigade, Ansar al-Sharia, Islamic State and al Qaeda.

Germany's ostensibly state-focused push for peace, although in

1 This chapter solely reflects the views of the author and not the Royal Military Academy Sandhurst, the UK Ministry of Defence or the British Army.

2 Mary Kaldor, *New and Old Wars: Organized Violence in a Global Era* (Cambridge: Polity Press, 2006), 1–2.

accordance with the strategy proposed by the UN-recognised Government of Libya (GNA), attracted criticism as well as plaudits.[3] Yet the optics, however poor, were also misleading. Haftar was in Berlin, if not at the gathering. As the participants knew well, Libya had long become a 'kingdom of militias', variously manipulated, supported and opposed by powerful regional and international forces.[4] Indeed, in a phenomenon observed in all but 1 of the 27 armed conflicts ongoing in 2018, the *dramatis personae* in this turbulent state are multiple groups of Violent Non-State Actors (VSNAs) pitted against, and in some cases cooperating with, a variety of state agents.[5]

The Berlin Conference thus provides a useful entry point to this volume. By 2020, Libya was approaching its tenth year of conflict. What had begun as a non-international armed conflict (NIAC) rapidly became a mixed NIAC/international armed conflict (IAC) comprising multiple overlapping, evolving conflicts, and a variety of actors – states acting against or in support of other entities, as well as large numbers of violent political groups of various hues.[6] More than this, however, Libya has also constituted a key destination and transit area for illicit flows, including the trafficking of goods (drugs, arms and counterfeit products) and people by violent organised criminal networks.[7] This constellation of actors makes

---

3 'Libya: Turning the Berlin Conference's Words into Action', *International Crisis Group*, 22 January 2020, available at https://www.crisisgroup.org/middle-east-north-africa/north-africa/libya/turning-berlin-conferences-words-action; Jeffrey Feltman, 'The Berlin Conference on Libya: Will Hypocrisy Undermine Results?' (Washington D.C.: Brookings Institution, 21 January 2020), available at https://www.brookings.edu/blog/order-from-chaos/2020/01/21/the-berlin-conference-on-libya-will-hypocrisy-undermine-results/; 'Can Europe, caught scrapping or napping, mend its credibility at Berlin talks on Libya?', *France24*, 18 January, 2020, available at https://www.france24.com/en/20200118-can-europe-caught-scrapping-or-napping-mend-its-credibility-at-berlin-talks-on-libya.

4 Jason Pack, *Kingdom of Militias: Libya's Second War of Post-Qadhafi Succession* (Milan: Instituto Per Gli Studi Di Politica Internazionale, 31 May 2019), available at https://www.ispionline.it/it/pubblicazione/kingdom-militias-libyas-second-war-post-qadhafi-succession-23121.

5 *SIPRI Yearbook 2019: Armaments, Disarmament and International Security* (Oxford: OUP, 2019).

6 Conflict classification is a matter of considerable debate. Beyond the mixed IAC-NIAC classification, Libya has also been categorised as an internationalised NIAC and an internalised IAC. Most analysts agree that several categories existed at any one time since 2011. For more see Louise Arimatsu and Mohbuda Choudhury, *The Legal Classification of the Armed Conflicts in Syria, Yemen and Libya.* (London: Chatham House, March 2014); Marko Milanovic, 'How to Qualify the Armed Conflict in Libya?', *EJILTalk*, 1 September 2011, available at https://www.ejiltalk.org/how-to-qualify-the-armed-conflict-in-libya/; Dapo Akande, 'Classification of Armed Conflicts: Relevant Legal Concepts', in Elizabeth Wilmshurst (ed.), *International Law and the Classification of Conflicts* (Oxford: OUP, 2012).

7 *Libya: a growing hub for Criminal Economies and Terrorist Financing in the Trans-Sahara, Policy Brief* (Geneva: The Global Initiative Against Globalized Transnational Organized Crime, 11 May 2015), available at https://globalinitiative.net/wp-content/uploads/2015/05/2015-1.pdf.

Libya instructive, even if it does not make it remarkable. Similar dynamics may be observed in Syria, Yemen, Iraq, the Democratic Republic of the Congo (DRC), the Central African Republic (CAR) and elsewhere. Each and all of these multifaceted conflicts can illuminate the discourse on VNSAs by offering insights into the considerable complexity of such actors and the challenges they pose.

In keeping with the aims of the *Sandhurst Trends in International Conflict* project, *Violent Non-State Actors in Modern Conflict* seeks to sharpen the focus of this discourse, more effectively delineate some of its core components, and provide impetus for further reflection and research. Incorporating theoretical, empirical and practitioner approaches, the volume seeks to critically engage with and offer policy-relevant insights into this complex and challenging phenomenon, thus providing a timely, multidisciplinary contribution to the growing literature on the subject.

The remainder of this Introduction unfolds in three substantive parts. It proceeds by highlighting several salient features of the debate surrounding VNSAs. Thereafter, it outlines and examines briefly the themes that feature prominently in this volume and which can be traced through its wide-ranging contributions. These can be grouped around four interrelated questions: how are VNSAs defined and categorised? How is VNSA behaviour evaluated? What kinds of relationships exist within VNSAs and between VNSAs and states? How should VNSAs be engaged and if necessary countered? A final section contains a synopsis of the chapters.

Neither Novel Nor New?

VNSAs have a long pedigree. State monopoly on the use of force was never as dominant as it appeared.[8] The perpetration of violence by groups lacking the authority to cause physical harm (whether obtained from God, the King or the State) is centuries old, as is the use of 'proxy' violence *in*, *for* and sometimes *by* states, which have tolerated, acquiesced in or even invited some form of cooperation with non-state actors (NSAs). In some respects, therefore, VNSAs are neither novel nor new.[9]

---

8 Phil Williams, *Violent Non-State Actors and National and International Security* (Zurich: International Relations and Security Network, ETH Zürich, 2008), 4.

9 Ulrich Schneckener, *Spoilers or Governance Actors? Engaging Armed Non-State Groups in Areas of Limited Statehood*, SFB-Governance Working Paper Series No. 21 (Berlin: SFB-Governance, October 2009), 6.

Nevertheless, if not new, per se, there is a *newness* about VNSAs, as evidenced by the academic and practitioner literature on the subject that has emerged over the last couple of decades.[10] Prominent here is the work of Mary Kaldor, who argued compellingly more than a decade before the onset of the Libyan conflict that 'old' war – classically conceived as centralised, territorialised interstate violence – was becoming anachronistic.[11] In its place she discerned a new type of organized violence. This 'new wars' thesis posited

> a blurring of the distinctions between war (usually defined as violence between states or organized political groups for political motivates), organized crime (violence undertaken by privately organized groups for private purposes, usually financial gain) and large-scale violations of human rights (violence undertaken by states or politically organized groups against individuals).[12]

Kaldor's *New and Old Wars* was an early, eloquent explication of a paradigm shift perceived in the explosion of multifaceted conflicts in the 1990s in Europe and elsewhere.[13] As the Stockholm International Peace Research Institute (SIPRI) reiterated in 2019, '[M]ost contemporary armed conflicts involve a combination of regular armies, militias and armed civilians. Fighting rarely occurs on well-defined battlefields and is often intermittent with a wide range of intensities and brief ceasefires'.[14] Furthermore, as

10 For a sample of the extensive VNSA literature see E. Chenoweth and A. Lawrence (eds) *Rethinking violence: states and non-state actors in conflict* (Cambridge, Massachusetts: MIT Press, 2010), Natasha Ezrow, *Global Politics and Violent Non-State Actors.* (London: Sage, 2017), Isabelle Duyvesteyn and Jan Angstrom (eds), *Rethinking the Nature of War* (Abington, Oxford: Routledge, 2005), Jason Bartolomei, Troy Thomas and William Casebeer, *Modelling Violent Non-State Actors: A Summary of Concepts and Methods*, Institute for Information Technology Applications, Research Publication 4 (Colorado: Institute for Information Technology Applications, United States Air Force Academy, November 2004), Paul Williams, *Violent non-state actors and national and international security* (Zurich: International Relations and Security Network, 2008), Robert Bunker, 'Defeating Violent Nonstate Actors', *Parameters* 43:4 (2013), Ulrich Schneckener, 'Fragile Statehood, Armed Non-State Actors and Security Governance', in A. Bryden and M. Caparini, *Private Actors and Security Governance* (Münster: LIT Verlag Münster), Mark Duffield, 'Post-modern conflict: Warlords, post-adjustment states and private protection', *Civil Wars* 1:1 (1998).

11 Kaldor, *New and Old Wars*, 17. See also Martin van Creveld, *The Transformation of War: The Most Radical Reinterpretation of Armed Conflict since Clausewitz* (New York: Free Press, 1991).

12 Kaldor, *New and Old Wars*, 2.

13 Development, Concepts and Doctrine Centre, *Global Strategic Trends: the future starts today*, 6th edition (Shrivenham: Development, Concepts and Doctrine Centre, 2018), 3. See also John Mueller, *Remnants of War* (USA: Cornell University Press, 2004). For a critique Mats Berdal, 'The 'New Wars' thesis revisited', in Hew Strachan and Sibylle Scheipers (eds), *The Changing Character of War* (Oxford: OUP, 2011).

14 *SIPRI Yearbook 2019.*

Briscoe has reminded us, some of the most lethal episodes of armed violence take place in countries that do not suffer from conflict, as conventionally defined.[15] The politico-criminal violence in Mexico is a case in point. As may be readily observed, therefore, contemporary VNSAs operate in an environment that is increasingly characterised by a pluralisation of irregular, non-conventional and privatised violence, often outside situations of armed conflict.[16] The antithesis of binary, such violence is not easily delineated.

Exploring the Paradigm Shift

What's New?

In considering further the implications of this new paradigm, including the putative polity shift from 'statehood to non-statehood', a useful starting point is to reflect upon the numbers.[17] Statistically, VNSAs appear to be more active than at any time in recent history. Since 2014, violent encounters (armed conflict and/or organised violence) solely between NSAs have been consistently higher than at any time since recording began in 1989. In 2019, 78 were recorded; prior to 2014, the highest number was 47 (recorded in 2001).[18] VNSAs are also more numerous. Even allowing for name duplication and the inclusion of states, state bodies and organisations such as NATO, the *Upsala Conflict Data Program Conflict Actor List* for 2017 contains some 6729 violent actors.[19] This trend is also reflected in the numbers of VNSAs involved in each 'individual' conflict. Although differentiation is not without its difficulties, the number of armed groups active per conflict seems to have increased, from an average of 8 in each of the intrastate conflicts extant in 1950 to 14 in the conflicts of 2010. In Mali, for instance, more than a dozen militias, separatists and jihadi groups have

---

15 Ivan Briscoe, *Non-conventional armed violence and non-state actors: challenges for mediation and humanitarian action.* (Norwegian Peacebuilding Resource Centre, May 2013), 2.

16 For more see Rafael Duarte Villa and Marilia Carolina Souza Pimenta, 'Violent non-State Actors and New Forms of Governance: Exploring the Colombian and Venezuelan Border Zone', *Journal of Human Security* 15:1 (2019).

17 For more see Ersel Aydinli, 'Assessing violent nonstate actorness in global politics: a framework for analysis', *Cambridge Review of International Affairs* 28:3 (2015).

18 For more see Upsala Conflict Data Program, available at https://ucdp.uu.se.

19 Upsala Conflict Data Program Actor List Dataset, available at https://ucdp.uu.se/downloads/.

been identified.[20] Meanwhile, in Syria, analysts have recorded over 1000 separate militias.[21] Libya's violence has involved as many as 2000 groups.[22] Significantly, contemporary VNSAs are also increasingly dynamic and fluid, 'merging and blending', forming alliances and sharing skills.[23] The dissemination of technical expertise, as seen in the exchange of bomb-making techniques between Hezbollah and al Qaeda, is an earlier example of this salient trend.[24]

A Coat of Many Colours?

This *newness* also has a qualitative difference. For one thing, VSNAs appear to come in a great many shapes and sizes: tribal and ethnic groups, pirates, warlords, gangs, terrorists, militias, insurgents, guerrilla fighters, rebels, private security companies and transnational criminal organisations, including narco-organisations. As these categories are rarely mutually exclusive, groups may play many roles concurrently. VNSAs also vary in character, motivation, goals, means, tactics, and impact. As one senior British General remarked, the forces he observed in Syria and Iraq circa 2016 amounted to 'a kaleidoscope of multifaceted organisations' exhibiting 'a kaleidoscope of loyalties, interests and objectives, some of which converge and some of which are distinct'.[25] Such diversity is perhaps one reason for the claim that the 'fundamental change' occasioned by the rise of VNSAs has been 'under-appreciated as a global phenomenon'.[26]

International Law and VNSAs

Beyond this, VNSAs are also increasingly part of the political and occasionally legal 'conversation', featuring in conflict-related and

---

20 Such as Mouvement National de Libération de l'Azawad (MNLA), Qaeda in the Islamic Maghreb (AQIM) and Ansar al Din. For more see Andrew Lebovic, *Mapping Armed Groups in Mali and the Sahel* (European Council on Foreign Relations, May 2019, available at https://www.ecfr.eu/mena/sahel_mapping#introduction; Sergei Boeke, 'Al Qaeda in the Islamic Maghreb: Terrorism, insurgency, or organized crime?', *Small Wars and Insurgencies* 25:5 (2016), 915, 918.

21 *SIPRI Yearbook 2018*, 18.

22 *SIPRI Yearbook 2018*, 18.

23 Robert J. Bunker, 'Defeating Violent Nonstate Actors', *Parameters* 43:4 (Winter 2013/14), 59.

24 For more see Michael C. Horowitz, 'Nonstate Actors and the Diffusion of Innovations: The Case of Suicide Terrorism', *International Organization* 64:1 (2010).

25 Lieutenant General Sir Mark Carlton-Smith, 'Testimony to the House of Commons Defence Select Committee on UK Military Operations in Syria and Iraq', 16 September 2016, available at https://publications.parliament.uk/pa/cm201617/cmselect/cmdfence/106/10602.htm.

26 Williams, *Violent Non-State Actors and National and International Security*, 4.

humanitarian discourse to a greater extent than before. They are cited frequently in UN Security Council (UNSC) Resolutions and Resolutions passed by the UN General Assembly (UNGA), the UN Human Rights Council (HRC) and other Treaty-based bodies, characterised, on occasion, as a threat to international peace and security.[27]

A growing number of International Treaties and Organisations exhort VNSAs to respect international norms and rules.[28] Under the *ius ad bellum* (pertaining to the threat and use of force), VNSAs are generally bound only insofar as they are under the 'effective control' of a state, with the onus being on the state to ensure compliance.[29] Made by states mostly for states, the *ius ad bellum* has little to say about sub-state groups. One notable source of contention therefore is whether a state may legally attack an NSA in self-defence. Although many legal scholars and governments broadly accept that force may be used (providing certain conditions are met), significant controversy remains over *how*, *when* and *where*.[30]

Elsewhere in international law, VSNAs are explicitly obligated to respect a growing number of international norms and rules. This can be seen clearly in International Humanitarian Law (IHL) and also *jus cogens* breaches of international law, such as genocide.[31] VNSAs are widely considered bound by the rules of the *ius jus in bello* in a NIAC, the argument being that they are bound by the law applicable within their state, of which IHL is a part.[32] As all states are parties to the Geneva Conventions, Common

27 See United Nations Security Council Resolution 2249 as an example. VNSAs have also been referenced in official statements produced by UN Special Procedures and Commissions, Inquiries and Fact-Finding Missions.
28 For more see, *Human Rights Obligations of Non-State Actors: An Exploration of the Practice of the UN Human Rights Council*, Academy In-Brief No. 7 (December 2016); A/71/353 *Outcome of the World Humanitarian Summit, Report of the Secretary General*, 23 August 2016 6/22.
29 Vladyslav Lanovoy, 'The Use of Force by Non-State Actors and the Limits of Attribution of Conduct', *The European Journal of International Law* 28:2 (2017), 563.
30 For a detailed overview see Donette Murray, 'Flawed and Unnecessary: the 'Unwilling or Unable' Doctrine Pertaining to States' Use of Force in Self-Defence Against Non-State Actors', *Hague Yearbook of International Law* 30 (2019); Noam Lubell, 'Fragmented Wars: Multi-Territorial Military Operations Against Armed Groups', in Christopher M. Ford and Winston S. Williams, *Complex Battlespaces: The Law of Armed Conflict and the Dynamics of Modern Warfare* (Oxford: OUP, 2019), 4–8.
31 The first UN Syria Commission reported that, 'at a minimum, human rights obligations constituting peremptory international law (*ius cogens*) bind States, individuals and non-State collective entities, including armed groups. Acts violating *ius cogens* – for instance, torture or enforced disappearances – can never be justified'. UN DOC A/HRC/19/69 (2012), *Report of the independent international commission of inquiry on the Syrian Arab Republic*, 20.
32 This is the positivist argument. As all states are signatories to the 1949 Geneva Conventions, the transnational character of VNSAs is immaterial.

Article 3 and Customary International Law (CIL) apply.[33] International Human Rights Law (IHRL), which has increasingly been applied in conflict (for instance in connection with occupation and detention) is also often cited as having some relevance to such groups.[34] And, since the International Criminal Court (ICC) became operational in 2002, individual VNSAs may face criminal charges under International Criminal Law (ICL) in relation to the most egregious international crimes, including crimes against humanity and war crimes.[35] In March 2012, Thomas Lubanga Dyilo, a Congolese warlord accused of war crimes, became the first person to be convicted at the ICC.

VNSAs may not possess 'active' legal personality – historically the preserve of states and some international organisations. Yet, neither are they entirely 'passive' subjects of international law, even if their precise 'place' in this rules-based system remains a matter of ongoing debate.[36] VNSAs can (and sometimes do) reject international law for ideological or political reasons, but are considered bound, albeit within certain limitations, regardless.

Themes in VNSAs

What's in a Name? Thinking about Definitions

Defined primarily by what it is *not* the term, VNSA covers a multitude of sins. Williams defines them as entities that challenge the state's 'monopoly

33 This is the positivist argument. For further discussion see Daragh Murray, 'How International Humanitarian Law Treaties Bind Non-State Armed Groups', *Journal of Conflict and Security Law* 20:1 (2015). Additional Protocol II may also apply if this has been ratified by the State in whose territory the NIAC occurs.

34 Specifically, regarding States, IHRL obligations are said to apply extraterritorially and in connection with detention. Regarding armed groups, CEDAW General Recommendation No. 30 notes that while non-state actors cannot be parties to international human rights treaties, in certain circumstances they are obliged to respect IHRL. *CEDAW General Recommendation No. 30 on women in conflict prevention, in conflict and post-conflict situations*, CEDAW C/GC/30 (18 October 2013). For a detailed account of this debate see Katherine Fortin, *The Accountability of Armed Groups Under Human Rights Law* (Oxford: OUP, 2017).

35 For more see the ICC's list of cases, available at https://www.icc-cpi.int/cases.

36 For a detailed overview see Donette Murray, 'Flawed and Unnecessary: the 'Unwilling or Unable' Doctrine Pertaining to States' Use of Force in Self-Defence Against Non-State Actors', *Hague Yearbook of International Law* 30 (2019); Noam Lubell, 'Fragmented Wars: Multi-Territorial Military Operations Against Armed Groups', in Christopher M. Ford and Winston S. Williams, *Complex Battlespaces: The Law of Armed Conflict and the Dynamics of Modern Warfare* (Oxford: OUP, 2019), 4–8.

on the use of violence within a specified geographical territory'.[37] Mandel, in contrast, emphasises both their 'relatively autonomous' nature and their 'significant and sustained coercive capabilities for organized violence'.[38] Another view, as seen in Bunker's definition, focuses on character, describing VNSAs as actors who 'exist along a threat continuum from that of common criminals to criminal soldiers'.[39]

The categorisation of VNSAs is similarly contested. Aydinli offers a typology based on autonomy, regeneration, representation and influence.[40] Alternatively, Schneckener's conceptual approach revolves around commonalities and differences. Typically, VNSAs share key features: the use of violence against unarmed civilians, the transnationalisation of groups and a preference for loose networks rather than hierarchical structures. They can also be differentiated, categorised in terms of aim: change *versus* status quo; location: territorial *versus* non-territorial; conduct: physical *versus* psychological use of violence; and motivation: political/ideological *versus* profit-driven motivation.[41]

These conceptual differences provoke several interesting questions. Should PMCs be categorised as VNSAs? Williams says not, arguing that, unlike most other VNSAs who operate outside the law, PMCs may 'reflect what might be termed the privatization of violence' but 'differ from other VNSAs in that they are, to a significant degree, regulated by states and generally act on behalf of states'.[42] Next, to what extent are VNSAs, labelled 'para-states' by some, actually state-like?[43] Relatedly, to what extent are VNSAs territorially bound? Do VNSAs always have a state link, however defined? And finally, how helpful is it to think of VNSAs as existing on a spectrum, with 'ordinary' criminal type groups towards one end and more political, para-military types towards the other? Islamic State (IS), for instance, has played many roles right across the spectrum.

---

37 Phil Williams, 'Here be Dragons: Dangerous spaces and international security' in A. L. Clunan and H. A. Trinkunas (eds), *Ungoverned Spaces: Alternatives to State Authority in an Era of Softened Sovereignty* (Stanford: Stanford Security Studies, 2010), 4.

38 Robert Mandel, *Global Security Upheaval: armed nonstate groups usurping state stability functions* (Palo Alto, California, United States: Stanford University Press, 2013), 42. A similar view is held by Claudia Hofmann and Ulrich Schneckener, 'Engaging non-state armed actors in state- and peace-building: options and strategies', *International Review of the Red Cross* 93:883 (September 2011), 2–3.

39 Bunker, 'Defeating Violent Nonstate Actors', 58.

40 See Aydinli, 'Assessing violent nonstate actorness in global politics'.

41 Schneckener, *Spoilers or Governance Actors?* 14–17.

42 Williams, *Violent Non-State Actors and National and International Security*, 9.

43 The Federation of American Scientists cited in Williams, *Violent Non-State Actors and National and International Security*, 4.

Evaluating VNSAs: Thinking about Behaviour

A second theme relates to the behaviour of VNSAs. Should this be evaluated from a normative, legal or even ethnographic perspective? Concerns about legitimation of illegal violence colours and shapes debates about VNSAs. At a very basic level, VNSAs appear to challenge states' monopoly on violence and are therefore viewed with considerable disquiet. Again, several questions are pertinent. Are VNSAs inherently illegitimate? Can any VNSAs be regarded as legitimate actors? Can VNSAs legitimate their actions and, if so, how?[44] A great deal hangs on these answers. As seen in discussions about IHL and political concepts such as the Responsibility to Protect (R2P), such anxieties are long-standing. States have frequently expressed concern that putative affirmation of VNSAs' *obligations* would also confer upon them *rights*.[45] Yet, as the 500 IHL–related commitments from more than 230 VNSAs brokered by the Swiss NGO Geneva Call shows, these entities can and do choose to adhere to international law.[46] Equally, so-called 'Robin Hood VNSAs' may fulfil a complex state-like role – delivering public goods and services – giving them *de facto* relevance within the polity and perhaps even some level of local legitimacy.[47] Labels plainly have policy and operational implications. Framing behaviour as legal/ illegal or legitimate/illegitimate creates red lines (as well as possibilities) for policymakers, both in terms of the nature of relations between the state and VNSAs, should cooperation or coordination be desirable or necessary, but also in terms of how action against such groups should be prosecuted if the relationship is adversarial.[48]

No VNSA is an Island: Thinking about Relationships

How VNSAs are defined and categorised, and how their behaviour is perceived and assessed, relate to a third issue. Whether opposed or

44 Isabelle Duyvesteyn, 'Rebels and Legitimacy: An Introduction', *Small Wars and Insurgencies* 28:4–5 (2017), 670.
45 For more see, *Violent Non-State Actors as Perpetrators and Enablers of Atrocity Crimes*, Policy Dialogue Brief (Indiana, USA: The Stanley Foundation, October 14–16, 2015).
46 For more see Ezequiel Heffes, 'Compliance with IHL by Non-State Armed Groups: Some Practical Reflections at the 70th Anniversary of the 1949 Geneva Conventions, *EJILTalk*, 21 August 2019, available at https://www.ejiltalk.org/compliance-with-ihl-by-non-state-armed-groups-some-practical-reflections-at-the-70th-anniversary-of-the-1949-geneva-conventions/.
47 Rajeev Chaudhry, 'Violent Non-State Actors: contours, challenges and consequences', *CLAWS Journal* (Winter 2013), 177.
48 Boeke, 'Al Qaeda in the Islamic Maghreb', 916.

co-opted, tolerated or ignored, viewed positively or negatively, VNSAs have a relational connection with states and those around them, including each other and the public.[49] This is an obvious but important point. For one thing, relationships expose states to political, military and legal risks, particularly when the line between 'acceptable' and 'non-acceptable' partners or proxies is blurred. The US policy of 'by, with and through' is one example. In its partnering with local anti-Assad forces in Syria, later abruptly abandoned, Washington chose to align with 100,000 fighters whom it judged to be acceptable (Syrian Democratic Forces (SDF), Provincial Internal Security Forces (PRISF) and the Internal Security Forces (InSF)) against those it had determined were not (including IS and al Qaeda).[50] In Ukraine, Russia's use of 'little green men' purportedly gave the Kremlin plausible deniability, although the consequences of this strategy remain hotly contested. Elsewhere, the activities of VNSAs have also overlapped with state governance.[51] In the border regions of Columbia and Venezuela, for instance, a type of fragile hybrid governance has emerged, raising questions about whether (or the extent to which) alternative and illegal VNSA-provided governance challenges the principles of statehood and the state-centric system.[52] And, as noted earlier, a complex web of inter-VNSA links has created a transnational network of inter-VNSA relationships, albeit one marked by loose, ad hoc and transient dynamics.

Dealing with VNSAs: Thinking about Responses

As highlighted above, *who* and *what* are key to determining responses to the challenges posed by VNSAs.[53] Not surprisingly, when contemplating the *how*, there is no one-size-fits-all.[54] Focusing on the more politico-military VNSAs, a large body of work has approached organised armed groups from

49 Ioana Emy Matesan and Ronit Berger, 'Blunders and Blame: How Armed Non-State Actors React to Their Mistakes', *Studies in Conflict and Terrorism* 45:1 (2017), 376–398.
50 *Syria: Turkish Incursion and Conflict Status* (Washington D.C.: Congressional Research Service Report, October 16 2019), 8-9 available at https://fas.org/sgp/crs/mideast/incursion.pdf.
51 Villa and Souza Pimenta, 'Violent Non-State Actors and New Forms of Governance'.
52 See Villa and Souza Pimenta, 'Violent non-State Actors and New Forms of Governance' and Aydinli, 'Assessing violent nonstate actorness in global politics', 439.
53 As Idler and Forest have remarked, 'Only by understanding the complex relationships that these VNSAs have with each other can we identify entry points that help dismantling the complementarity of their governance, which, in turn, deprives VNSAs of their social base, rooted in their legitimization by the local population'. Annette Idler and James J. F. Forest, 'Behavioral Patterns among (Violent) Non-State Actors: A Study of Complementary Governance', *International Journal of Security and Governance* 4(1):2, 13.
54 Schneckener, *Spoilers or Governance Actors?*

a counter-terrorism or counter-insurgency perspective. Perhaps viewing the phenomenon more broadly, others have pressed IR theory into service, utilising Realist, Institutionalist and Constructivist approaches (advocating respectively the power of *force*, the power of *bargaining* and the power of *persuasion*) to create a bespoke response appropriate to the group or groups in question.[55] Alternatively, some have privileged so-called 'spoiler management' strategies[56] or argued that that states must adopt a 'start-up mentality', acquiring the ability 'to respond creatively to the challenges and opportunities of new technologies'.[57] Notwithstanding these different perspectives, most would also agree that context is critical and VNSAs rarely exist in true isolation.

Adopting a narrow focus is likely to produce only limited success. Military and policing responses are important. Countering VNSAs may actually take the form of engagement, particularly in the humanitarian space, as the International Committee of the Red Cross (ICRC) has shown with its development of a framework for engagement with organised armed groups. Countering in its broadest sense is determined by issues of timing and opportunity, the ability to find pressure points and exert leverage, often evolving as circumstances change, particularly as conflict segues into peace-making and beyond.[58] With such a variety of groups in play, motivations, goals, character, capacity and behaviour differentiate between VNSAs and shape the extent to which a very narrow military option can be complemented or even supplanted by a political approach. Moral concerns may also be apposite. Relevant too are prosaic challenges – internal fragmentation within groups frustrates communication, impinging upon efforts to establish dialogue. Additionally, with VNSAs of a mixed identity, the 'traditional' goal of achieving a 'political peace' may be at best of only partial interest to some of the actors involved.[59] Finally, it is worth considering what role (if any) such groups should have in various settings: stabilisation efforts, building peace and providing for public

55 Hofmann and Schneckener, 'Engaging non-state armed actors in state- and peace-building: options and strategies', 6.

56 See Stephen J. Stedman, 'Spoiler problems in peace processes', *International Security* 22:2 (1997).

57 Daveed Gartenstein-Ross and Natheniel Barr, 'The Lean Terrorist Cell: How Startup Companies and Violent Non-State Actors Are Challenging the Old World Order', *Georgetown Journal of International Affairs* 17:2 (2016), 32.

58 Hofmann and Schenckener, 'Engaging non-state armed actors in state- and peace-building: options and strategies', 3.

59 For more see Heffes, 'Compliance with IHL by Non-State Armed Groups'.

goods, including security. As Idler and Forest have argued, in some cases the complementary governance provided by VNSAs can result in higher levels of security than the government can provide.[60]

A last point is that different approaches carry different risks as well as opportunities. Law, for instance, enables the use of force but also constrains it. Blacklisting may help to build a case against VNSAs, but makes it more difficult to change direction if this strategy fails.[61] Legal sanctions and injunctions may attract states' obligations whilst restricting their capacity to engage.[62] International law is also unsettled in places, meaning that what states can and cannot do to counter such groups is not always clear. Governments' use of force against NSAs and the extent to which VNSAs can be held accountable for their actions, particularly where violence has been dealt to civilians, are particularly contentious. The same is true for the *ius in bello*. Many states have expressed concerns about the ICRC's 2009 *Interpretative Guidance on the Notion of Direct Participation in Hostilities* and, specifically, that the putative protections envisaged for violent civilians privileges them over their state adversaries.[63] A lack of awareness of legal concepts and specific rules in IHL hampers a rules-based approach; the vexed question of detention of VNSAs within a NIAC is similarly troublesome.[64] In countering serious organised violence, therefore, legal questions, including how conflicts are classified and which law is applicable – IHL, IHRL or ICL – form an important part the discourse.[65]

60 Idler and Forest, 'Behavioural Patterns among (Violent) Non-State Actors', 4.

61 *Violent Non-State Actors as Perpetrators and Enablers of Atrocity Crimes*, 3.

62 A number of UNSC resolutions are concerned with countering NSAs and/or specifically outline States' obligations to do so, including UNSC 1540, UNSC 2249 and UNSC 2325.

63 For more see Ryder McKeown, 'Legal Asymmetries in Asymmetric War', *Review of International Studies* 41:1 (January 2015), 126.

64 For more see Heffes, 'Compliance with IHL by Non-State Armed Groups'; Grant Davies, 'Revolving Doors and Human Rights: Detention in Internationalised Non-International Armed Conflict', in David Brown, Donette Murray, Malte Riemann, Norma Rossi, and Martin A. Smith, *War Amongst the People: Critical Assessments* (Havant: Howgate, 2019).

65 Marko Milanovic, 'Accounting for the Complexity of the Law Applicable to Modern Armed Conflicts', in Christopher M. Ford and Winston S. Williams, *Complex Battlespaces*, 49; McKeown, 'Legal Asymmetries in Asymmetric War', 129; Gloria Gaggioli (ed.), *ICRC Report of the Expert Meeting on The Use of Force in Armed Conflicts: Interplay Between the Conduct of Hostilities and Law Enforcement Paradigms* (Geneva: ICRC, 2013); Elizabeth Wilmshurst (ed.), *International Law and the Classification of Conflicts* (Oxford: OUP, 2012).

Synopsis of the Book

The four themes detailed in the section above are elaborated in the chapters that follow. In the first of three conceptual chapters exploring how we think about, understand and classify VNSAs, Caroline Varin considers the erosion of categories in a definitionally focused analysis. Drawing upon a number of African case studies, she argues that the differences between terrorists, warlords and rebels have progressively disappeared as they merge with one another, overlapping in terms of their ambitions to control territory and in their ability to challenge the sovereign state. This convergence over territory, a criterion that previously distinguished terrorists from rebels and warlords in Africa, makes it harder to usefully categorise VNSAs in modern conflict. Notwithstanding this convergence, Varin argues that differing political ambitions and methods remain pertinent: warlords remain fairly static actors, rebels seek to replace the state, while terrorist ambitions on the African continent transcend borders and politics. As a consequence, she concludes, effective counter-terrorist and counter-insurgency strategies must begin with a more precise understanding of the nature, ambitions and methods of the enemy. To this end, what is needed is a new criterion or criteria, other than the possession, governance or ambition to hold territory. Without greater conceptual granularity, and a recognition that political labelling of groups can be problematic, effective policymaking will be impeded.

In Chapter Two, Helene Olsen engages with the definitional and categorisation challenges raised by Varin using a different lens. Focusing on mercenaries, her contribution explores issues of legitimacy and relationships. She makes two substantive points. First, these VNSAs are not inherently illegitimate, but are rather constructed as such through distinct discursive practices that facilitate the conceptual understanding of mercenaries as illegitimate fighters. Mercenaries, she argues, are assigned illegitimacy through the repeated usage of a specific set of identifiable moral objections expressed in connection with their usage and have been most violently objected to and de-legitimised in situations where their presence on the battlefield has threatened prevalent conceptions of the ideal polity. Olsen's second claim is that the process of de-legitimisation provides valuable insights into the relationship between mercenaries, their employers and the surrounding polity. This analysis suggests that, by appreciating that resistance towards mercenaries and their active de-legitimation as fighters has been a common practice, the future utility of these actors can be discussed on a more informed, nuanced and operationally useful basis.

Similarly concerned with issues of definition and categorisation, Helle Malmvig takes aim at the misleading *NSA* prefix, arguing that it obscures how closely interwoven with state power, or state-like, VNSAs are. In Chapter Three, she sets about challenging prevalent assumptions about what a state *is* (a polity with exclusive territorial control and monopoly on violence) and, thereafter, about what VSNAs *do*. Regarding the latter, Malmvig examines whether such groups necessarily challenge and undermine 'stateness'. Using Syria as an example, she demonstrates that the boundary between state and non-state is 'far from a zero-sum relationship'. Extremely fluid and continuously contested, multiple intersecting nodes of authorities, actors and overlapping relationships make the dynamics complex and challenging for all those involved.

Building on Varin, Olsen and Malvig's conceptual foundations, four empirically rich case study-based contributions offer insights into the complex relationships between VNSAs and a variety of politically diverse states. In Chapter 4, Jack Watling's ethnographical account draws attention to the apparent shortcomings of international legal frameworks with regards to regulating the interactions between states and VNSAs. From his perspective, a gulf exists between international relations as described in law and what he relates as the reality on the ground. Key to his analysis is how best to shape the behaviour of VNSAs. Legitimacy, he argues – understood in terms of the legitimacy of external actors and IHL – must be grounded in local norms and customs. As such, states need to consider under what circumstances VNSAs can contribute to the development of IHL, since effective law is agreed by practitioners.

Approaching the theme of relationships from a strategic client-patron perspective, James Fargher's contribution examines the consequences of VNSA-state interaction. He argues in Chapter 5 that, by sparking an international conflict in Yemen, the Houthis facilitated the expansion of Emirati and Saudi naval influence in the region. In other words, their actions provided 'a key justification for Abu Dhabi and Riyadh to seize islands, to purchase naval bases, and to form strategic alliances with small local states'. This raises a number of interesting questions: did the Houthis (and their external backers, Iran) miscalculate or have they secured their politico-military aims? And, presumably having recognised the advantages accrued by the coalition forces ranged against them, how and to what extent has this particular VNSA-state alliance sought to reverse the putative gains made by their adversaries?

In Chapter 6, Marina Miron focuses on Russia's use of VNSAs in support of Bashir al-Assad's regime. Though not oblivious to the problems it has confronted in dealing with Kurdish proxies and the Wagner group, Miron argues that much seems to have gone right for Moscow in Syria. Not only did it utilise VNSAs in 'a very complex environment', but, significantly, President Putin's government also 'gained strategic outcomes' as a result of this strategy. Critically, it achieved its long-term goal of saving a key regional ally, whilst denying the West what it perceived as another regime change in one of its traditional spheres of influence. Furthermore, and in the short-term, the Kremlin also took action to augment its internal security by confronting the threat of Syrian jihadism 'at source'. Characterising this as 'adroit' use of two quite different violent armed groups in Syria, Miron concludes that the Russian experience offers lessons about how best to engage with VNSAs.

Abigail Watson concurs that the use of VNSAs *can* be beneficial, but deviates from Miron's conclusions, particularly in terms of any cost-benefit analysis. For Watson these relationships are deeply problematic. In Chapter 7, she notes that 'operational gains have convinced many that this type of engagement can allow governments to act in the face of credible threats abroad without asking too much of squeezed defence budgets and risk-averse publics and parliaments'. Yet, she warns too that policymakers must recognise the dangers involved. In many cases, the short-term tactical gains are far outweighed by long-term strategic losses. As Watson's case studies of Syria and Libya show, the long-term aims of those involved have diverged and will likely continue to do so, with serious consequences. Differing goals amongst the participants risk exacerbating political divides and prolonging the conflict. With this in mind, she warns that, if 'relationships with partners are not chosen and developed carefully, they risk exacerbating internal divisions and potentially undermining the long-term prospects for peace and stability'.

A third section complements the academic analysis, engaging with it from a practitioner's perspective. In their contributions, Julien Bastrup-Birk, Grant Davies and Matthew Walker – specialists with expertise in stabilisation, international law and military operations – reflect retrospectively upon the key practical challenges posed by VNSAs and assess current and future trends in relation to VNSAs in modern conflict. Bastrup-Birk argues that VNSAs pose an enduring challenge to those trying to counter them, not least because they possess particular advantages – agility, access to local political constituencies, knowledge of the ground and, increasingly, technology. He identifies several dynamic challenges: a rapidly evolving threat; 'competing' with alternative political structures; local knowledge and strategic patience; ungoverned spaces and border areas. In Bastrup-Birk's experience, there is no 'one-size-fits-all' or magic bullet(s). States and international organisations seeking to respond to this threat must be aware that future trends point to a migration of ideas and tactics, the use of technology as a force multiplier and a growing tendency to utilise VNSAs as proxies.

Grant Davies examines VNSA through a legal prism and begins by considering how VNSA's are defined in international law. Focusing on IHL, he highlights what the Geneva Conventions say, and do not say, about those who may be categorised thus. Davies also points out that, to the right of the 'bright red line' of armed conflict, a degree of certainty exists, based on a body of law and rules that have developed into a framework under which armed forces and states can operate, particularly in international conflict. However, to the left of this line, as Watling has also lamented, things are considerably more opaque. Such uncertainty, Davies warns, can and is exploited, not only by VNSAs, but also by states seeking an advantage against their opponents. To develop this further he identifies several key challenges: the problem of defining an 'armed attack' by VNSAs under *ius ad bellum*; attribution of VNSAs' actions to a state; and the legal challenges for states who cooperate with and provide assistance to VSNAs, particularly in connection with detention in NIAC. Combined, a paucity of law, particularly in NIAC, differing political interpretations about the existing legal framework and operational realities on the ground present substantial challenges in dealing with VNSAs.

In the third of the practitioner contributions, Matthew Walker considers VNSAs from a doctrinal and operational perspective. His first argument is a critical one: initially at least, intervening forces lacked any corporate memory to guide their actions. As such, UK forces and others had ground to make up. As Bastrup-Birk and Davies also observed, VNSAs have enjoyed considerable advantages. As he notes, 'operating through, and within, indoctrinated or aligned members of the wider community, VNSAs' capacity to take ownership of the operational and strategic narratives presents a significant practical challenge'. VNSAs 'vital ground', Walker suggests, is in the information domain. Only a co-ordinated 'whole of government' approach – one that exploits all diplomatic, economic, cultural and military levers of national power – will suffice. His final point concerns the need for military adaptation, perhaps even transformation.

Traditionally rooted in the physical dimension, in the future, niche military capabilities, such as the need to deliver information manoeuvre is likely to be as important as the coercive military hard power. Recognising and accommodating this will require considerable conceptual and practical adjustments.

In the concluding chapter, Norma Rossi and Malte Riemann reiterate the value of a multi-faceted perspective that draws on mixed methodologies and different epistemological approaches. In so doing, they demonstrate that 'analysis of VNSAs is crucial for developing improved and more nuanced conceptual, empirical, normative and policy-relevant understandings of contemporary forms of conflict and their strategic configurations in global politics'. Framing their contribution as a challenge to some of the inherent assumptions that structure debates on VNSAs, Rossi and Riemann make three key arguments. First, they amplify and explicate the critique made in these pages about the assumption of state-centrism. Second, and by extension, they argue that the growing role of VNSAs in conflict can be read as a symptomatic expression of a process leading to the multiplication of sovereign centres. This, in turn, contributes to a wider debate about whether (or the extent to which) sovereignty can be regarded as an exclusive feature of states. Third, in reflecting upon the policy implications of these claims, Rossi and Riemann demonstrate that it is possible to 'bridge the academic/ practitioner divide and through this expose crucial conundrums states face in dealing with VNSAs in contemporary conflict'.

References

Akande, D. 'Classification of Armed Conflicts: Relevant Legal Concepts', in Elizabeth Wilmshurst (ed.), *International Law and the Classification of Conflicts*. Oxford: OUP, 2012.

Arimatsu, L. and Choudhury, M. The Legal Classification of the Armed Conflicts in Syria, *Yemen and* Libya. London: Chatham House, March 2014.

Aydinli, E. 'Assessing violent nonstate actorness in global politics: a framework for analysis', *Cambridge Review of International Affairs* 28:3, 2015.

Bartolomei, J., Thomas, T. and Casebeer, W. *Modelling Violent Non-State Actors: A Summary of Concepts and Methods*, Institute for Information Technology Applications, Research Publication 4. Colorado: Institute for Information Technology Applications, United States Air Force Academy, November 2004.

Berdal, M. 'The 'New Wars' thesis revisited', in Hew Strachan and Sibylle Scheipers (eds), *The Changing Character of War*. Oxford: OUP, 2011.

Boeke, S. 'Al Qaeda in the Islamic Maghreb: Terrorism, insurgency, or organized crime?' *Small Wars and Insurgencies* 25:5, 2016.

Briscoe, I. *Non-conventional armed violence and non-state actors: challenges for mediation and humanitarian action*. Norwegian Peacebuilding Resource Centre, May 2013.

Brown, D., Murray, D., Riemann, M., Rossi, N. and Smith, M. A. *War Amongst the People: Critical Assessments*. Havant: Howgate, 2019.

Bunker, R. 'Defeating Violent Nonstate Actors', *Parameters* 43:4, 2013/4.

Carlton-Smith, Lt General Sir Mark. 'Testimony to the House of Commons Defence Select Committee on UK Military Operations in Syria and Iraq', 16 September 2016.

CEDAW. *General Recommendation No. 30 on women in conflict prevention, in conflict and post-conflict situations*, CEDAW C/GC/30, 18 October 2013.

Chaudhry, R. 'Violent Non-State Actors: contours, challenges and consequences', *CLAWS Journal*, Winter 2013.

Chenoweth, E. and Lawrence, A. (eds). *Rethinking violence: states and non-state actors in conflict*. Cambridge, Massachusetts: MIT Press, 2010.

Development, Concepts and Doctrine Centre. *Global Strategic Trends: the future starts today*, 6th edition. Shrivenham: Development, Concepts and Doctrine Centre, 2018.

Duffield, M. 'Post-modern conflict: Warlords, post-adjustment states and private protection', *Civil Wars* 1:1, 1998.

Duyvesteyn, I. 'Rebels and Legitimacy: An Introduction', *Small Wars and Insurgencies* 28:4 5, 2017.

Duyvesteyn, I. and Angstrom, J. (eds). *Rethinking the Nature of War*. Abington, Oxford: Routledge, 2005.

Ezrow, N. *Global Politics and Violent Non-State Actors*. London: Sage, 2017.

Ford, C. M and Williams, W. S. *Complex Battlespaces: The Law of Armed Conflict and the Dynamics of Modern Warfare*. Oxford: OUP, 2019.

Fortin, K. *The Accountability of Armed Groups Under Human Rights Law*. Oxford: OUP, 2017.

Gaggioli, G. (ed). *ICRC Report of the Expert Meeting on The Use of Force in Armed Conflicts: Interplay Between the Conduct of Hostilities and Law Enforcement Paradigms*. Geneva: ICRC, 2013.

Gartenstein-Ross, D. and Barr, N. 'The Lean Terrorist Cell: How Startup Companies and Violent Non-State Actors Are Challenging the Old World Order', *Georgetown Journal of International Affairs*, 17:2, 2016.

Heffes, E. 'Compliance with IHL by Non-State Armed Groups: Some Practical Reflections at the 70th Anniversary of the 1949 Geneva Conventions, *EJILTalk*, 21 August, 2019.

Hofmann, C. and Schneckener, U. 'Engaging non-state armed actors in state- and peace building: options and strategies', *International Review of the Red Cross* 93:883, September 2011.

Horowitz, M. C. 'Nonstate Actors and the Diffusion of Innovations: The Case of Suicide Terrorism', *International Organization* 64:1, 2010.

Human Rights Obligations of Non-State Actors: An Exploration of the Practice of the UN Human Rights Council, Academy In-Brief No. 7. December 2016.

Idler, A. and Forest, J. J. F. 'Behavioral Patterns among (Violent) Non-State Actors: A Study of Complementary Governance', *International Journal of Security and Governance* 4(1):2, 2015.

Kaldor, M. *New and Old Wars: Organized Violence in a Global Era.* Cambridge: Polity, 2006.

Lanovoy, V. 'The Use of Force by Non-State Actors and the Limits of Attribution of Conduct', *The European Journal of International Law* 28:2, 2017.

Libya: a growing hub for Criminal Economies and Terrorist Financing in the Trans-Sahara, Policy Brief. Geneva: The Global Initiative Against Globalized Transnational Organized Crime, 11 May 2015.

Mandel, R. *Global Security Upheaval: armed nonstate groups usurping state stability functions.* Palo Alto, California, United States: Stanford University Press, 2013.

Matesan, I. E. and Berger, R. 'Blunders and Blame: How Armed Non-State Actors React to Their Mistakes', *Studies in Conflict and Terrorism* 45:1, 2017.

McKeown, R. 'Legal Asymmetries in Asymmetric War', *Review of International Studies* 41:1, January 2015.

Milanovic, M. 'How to Qualify the Armed Conflict in Libya?', *EJILTalk!*, 1 September 2011.

Milanovic, M. 'Accounting for the Complexity of the Law Applicable to Modern Armed Conflicts', in Christopher M. Ford and Winston S. Williams, *Complex Battlespaces: The Law of Armed Conflict and the Dynamics of Modern Warfare*. Oxford: OUP, 2019.

Mueller, J. *Remnants of War*. Ithaca, NY, USA: Cornell University Press, 2004.

Murray, D. 'How International Humanitarian Law Treaties Bind Non-State Armed Groups', *Journal of Conflict and Security Law*, 20:1, 2015.

Murray, D. 'Flawed and Unnecessary: the 'Unwilling or Unable' Doctrine Pertaining to States' Use of Force in Self-Defence Against Non-State Actors', *Hague Yearbook of International Law* 30, December 2019.

Pack, J. *Kingdom of Militias: Libya's Second War of Post-Qadhafi Succession.* Milan: Instituto Per Gli Studi Di Politica Internazionale, 31 May 2019.

Schneckener, U. 'Fragile Statehood, Armed Non-State Actors and Security Governance', in A. Bryden and M. Caparini, *Private Actors and Security Governance*. Münster: LIT Verlag Münster.

Schneckener, U. *Spoilers or Governance Actors? Engaging Armed Non-State Groups in Areas of Limited Statehood*, SFB-Governance Working Paper Series No. 21. Berlin: SFB-Governance, October 2009.

SIPRI Yearbook 2019: Armaments, Disarmament and International Security. Oxford: OUP, 2019.

Stedman, S. J. 'Spoiler problems in peace processes', *International Security* 22:2, 1997.

Syria: Turkish Incursion and Conflict Status (Washington D.C.: Congressional Research Service Report, October 16 2019.

UN Doc A/71/353 Outcome of the World Humanitarian Summit, Report of the Secretary General, 23 August 2016.

UN Doc A/HRC/19/69, Report of the independent international commission of inquiry on the Syrian Arab Republic, 2012.
Upsala Conflict Data Program Actor List Dataset, available at https://ucdp.uu.se/downloads/.
Violent Non-State Actors as Perpetrators and Enablers of Atrocity Crimes, Policy Dialogue Brief. Indiana, USA: The Stanley Foundation, October 14-16, 2015.
Van Creveld, M. *The Transformation of War: The Most Radical Reinterpretation of Armed Conflict since Clausewitz*. New York: Free Press, 1991.
Villa, R. D. and Souza Pimenta, M. C. 'Violent non-State Actors and New Forms of Governance: Exploring the Colombian and Venezuelan Border Zone', *Journal of Human Security* 15:1, 2019.
Williams, P. 'Here be Dragons: Dangerous spaces and international security' in A. L. Clunan and H. A. Trinkunas (eds), *Ungoverned Spaces: Alternatives to State Authority in an Era of Softened Sovereignty*. Stanford: Stanford Security Studies, 2010.
Williams, P. *Violent Non-State Actors and National and International Security*. Zurich: International Relations and Security Network, ETH Zürich, 2008.

1

A CONVERGENCE OF TERRITORY

Warlords, Rebels and Terrorists in the Modern African State

Caroline Varin

Introduction

The issue of territory has long been contentious in international relations and international law. In international law, territory can only belong to the state, which remains the central actor in international relations. According to Oppenheim, 'a State without a territory is not possible',[1] a statement reinforced by Shaw who claims that 'statehood is inconceivable in the absence of a reasonably defined geographical base'.[2] This Western concept of statehood was imposed on Africa during colonialism and remains the main qualification for sovereignty in the international system. This is further entrenched in the 1963 Charter of the Organization of African Unity (OAU), which crystalizes the OAU's commitment to protecting the trinity of 'sovereignty, territorial integrity and independence'.[3]

Since the end of colonialism, however, the territorial sovereignty of African states has been repeatedly and successfully challenged by rebels and warlords, who have taken over huge swaths of land, attacked and occupied

---

1 Lassa Oppenheim, *International Law* Vol – 1. 8 (1955), 451-52.

2 Malcolm Shaw. 'Territory in International Law'. *Netherlands Yearbook of International Law,* 13 (1982), 61-91.

3 Charter of the Organization of African Unity (1963), 3.

cities and even overthrown the sitting government. This has been possible largely as a result of weak state institutions that have focused on keeping control of the cities, leaving the space in-between vulnerable to non-state actors:[4] Where there is greater state capacity to maintain control over the entire territory, insurgents are less able to challenge territorial sovereignty, remain underground and revert to terrorist tactics to undermine the state.

Historically, challenges to the sovereignty of African states have been overwhelmingly in the form of the territorial ambitions of rebel groups and local warlords, often supported by foreign actors. Recently however, terrorist groups have also aspired to hold territory in Africa. This Violent Non-State Actor (VNSA) has traditionally been characterized by its politically driven terror tactics within cities,[5] but territory had not previously featured in most definitions of terrorists. Indeed, in 2012, De la Calle and Sanchez-Cuenca specifically described terrorist groups as 'those, unlike guerrillas, (who) act underground, without de facto territorial control'.[6] They further emphasize that 'confusion between terrorist groups and rural guerrillas rarely emerged' as guerrillas 'challenge the state's sovereignty, by segmenting of fragmenting it' whereas terrorists do not.[7]

This distinctive pattern of behaviour has shifted, starting with Al-Qaeda in the Islamic Maghreb's (AQIM) successful occupation of northern Mali in 2012-2013; although it should be noted that Al-Shabaab's ambitions also took on a territorial dimension in Somalia as early as 2006.[8] All four major terrorist groups in Africa have explicit territorial ambitions and were able to rapidly 'conquer' and control huge swathes of land in the early twenty-first century. Having succeeded in their quest in 2012, AQIM revealed that such ambitions were a realistic objective for terrorist groups based in states with a weak government and limited military capacity that fell beyond the interests of the Western world. Boko Haram's expansion into north-eastern Nigeria and ISIS' establishment in Libya also reveal an ongoing trend for this VNSA.

This chapter draws on the case studies from Varin and Abubakar's 2017 edited volume *Violent Non-State Actors in Africa: Terrorists, Rebels and*

---

4 Laura Dugan, Gary LaFree, Luis De La Calle, Ignacio Sánchez-Cuenca, Victor Asal, Michael G. Findley, and Joseph K. Young. 'Killing Civilians or Holding Territory? How to Think about Terrorism'. *International Studies Review* 14, no. 3 (2012): 475-97.

5 Martha Crenshaw, 'The Causes of Terrorism'. *Comparative Politics* 13, no. 4 (1981): 379-99.

6 Dugan, 'Killing Civilians or Holding Territory? How to Think about Terrorism', 481.

7 Dugan, 'Killing Civilians or Holding Territory? How to Think about Terrorism', 481.

8 Bohumil Doboš, 'Shapeshifter of Somalia: Evolution of the Political Territoriality of Al-Shabaab', *Small Wars & Insurgencies*, 27:5 (2016): 937-957.

Warlords to demonstrate the convergence of terrorists, rebels and warlords in their ambitions to pursue 'political territoriality'[9] and their ability to successfully hold and control this territory over a certain period of time. The chapter explains that differences between terrorists, warlords and rebels have progressively disappeared in an African context and argues that it has become increasingly difficult to distinguish between these VNSAs as they merge with one another, overlapping in their ambitions to control territory and their ability to challenge the sovereign state. It begins by exploring the concept of territoriality and sovereignty in an African context before examining in turn the differences between warlords, rebels and terrorists with regards to their territorial ambitions, drawing examples from the cases developed in Varin and Abubakar's volume.

Sovereignty and VNSAs in an African Context

The literature on VNSAs encompasses a variety of players, including terrorists, rebels, warlords and pirates, but also insurgents and mercenaries. Definitions of these groups have often been highjacked by political actors to fulfil their own ambitions, and there is also no consensus in the academic or policy world to describe and distinguish them from each other. For the purpose of this study, VNSAs have been defined in their broadest terms as *a group of civilians who are organised together in the illegitimate exercise of violence.*[10] They differ in size, characteristics, organisation structure, motive, ideology and methods, but have in common that they have all resorted to the illegitimate use of violence, which puts them in direct conflict with the state who claims a monopoly of violence.[11] Although these same actors are sometimes in the employ of or act with the strategic encouragement of the government, the groups involved in this study are by definition in juxtaposition with the state as they specifically seek to undermine it by means of violence and by taking over territory, the most sacred property of the contemporary state. Diane Davis describes non-state armed actors

9 Doboš, 'Shapeshifter of Somalia: Evolution of the Political Territoriality of Al-Shabaab'.
10 Caroline Varin, Abubakar Dauda, 'Violent Non-State Actors in Africa: Terrorists, Rebels and Warlords' (Palgrave 2017), 5.
11 Dugan, 'Killing Civilians or Holding Territory? How to Think about Terrorism', 481.

as 'flying under the radar screen (if not catapulting over) the broadly cast domains of governance and sovereignty associated with the nation-state'.[12]

The question of state legitimacy is a tenuous one, especially in the case of Africa, where authority has frequently been imposed externally, falling foul of Marx and Tilly's theories on state-making and sovereignty.[13] The 54 African states are the political result of the power relations of nineteenth century Europe, with legitimacy afforded the territorial government by means of external recognition, rather than built upon a social contract between the people and their rulers.[14] Weber's definition of sovereignty in terms of 'the monopoly over the legitimate use of force within a given territory'[15] certainly does not apply to most African states;[16] however, by virtue of being internationally accepted as the sole representative of its population, the state can legitimately label and denounce violence undertaken by any non-state actor as unlawful. Indeed, Robert Jackson and Carl Rosberg make this distinction between the state's *de facto* and *de jure* statehood.[17] Whereas the state and a VNSA may perform the exact same level of terror and destruction on a population, the former is legitimised (if not legitimate) and the latter is criminalised by the state and the international community.

While there are many categories of VNSAs, this chapter adopts three specific types of actor: rebels, warlords and terrorists, with the objective of exploring the growing convergence around their territorial ambitions. The distinction between rebels, warlords and terrorists is sometimes clear; at other times it is difficult to make. Although these VNSAs are categorized separately, they can change in the course of their existence, creating new ties, exploiting new opportunities, and therefore changing their objectives, membership and operations. This is facilitated in an era of globalization where networks are transnational, therefore sustaining each other for a much longer period. As a result, groups that exploit local resources can trade across borders and further abroad, taking advantage of new developments in communications and transportation. Terrorists, rebels and warlords are

12 Diane E. Davis, 'Non-State Armed Actors, New Imagined Communities, and Shifting Patterns of Sovereignty and Insecurity in the Modern World', *Contemporary Security Policy*, 30:2 (2009), 221–245.
13 Charles Tilly, Capital, Coercion, and European States, AD 990-1992 (Cambridge: Basil Blackwell, 1990.
14 Pierre Englebert, *State Legitimacy and Development in Africa*. (Lynne Rienner, 2002), 96.
15 Max Weber, *The Protestant Ethic and the Spirit of Capitalism*. (New York: Scibner's Sons, 1958).
16 Robert H. Jackson and Carl G. Rosberg, 'Why Africa's Weak States Persist: The Empirical and the Juridical in Statehood' *World Politics*, Vol. 35, No. 1 (1982).
17 Jackson 'Why Africa's Weak States Persist: The Empirical and the Juridical in Statehood'.

known to have engaged in and rely on extensive criminal activities locally and internationally with other illegitimate actors. Boko Haram is one such example. It has been called a terrorist group, a criminal organisation, an insurgency and a rebel group. Under Abu Shekau, while he was in control of a large part of the northeast of Nigeria in 2015, one might even have called him a warlord.[18]

Rebels and terrorists often share a common goal of bringing about political change. The difference between the two may be measured at first by the scale of the group and the organization and scope of the violence.[19] Who they target is also of importance. While rebel groups mainly attack the state and its representatives such as the police, armed forces and the population in the immediate vicinity of their operational base, terrorists have tended to include foreign and especially western interests as their top targets.[20] Rebels are supported by and are able to mobilize a significant portion of the local population, making them first and foremost a domestic actor. The size and support base of the group are crucial for terrorism expert Bruce Hoffman, who explains that rebels refer to a 'numerically larger group of armed individuals' than are present in terrorist groups.[21] Bell and Evans define rebels as 'overtly nationalist, spatially contained and classically defensive...designed to liberate a homeland' from a tyrant or an unacceptable way of life; by contrast 'the terrorist is almost always in a minority..., they rarely have majority support, let alone some alternative blueprint as to how their vision of a better world will work'.[22] Terrorists rarely bring about the political change they aspire to, although they can cause a lot of damage in their attempt to do so.

This chapter argues that, notwithstanding other similarities, terrorists, rebels and warlords in Africa have converged in their mission to exercise political territoriality. Robert Sack defines territoriality as 'the attempt by an individual or group to affect, influence, or control people, phenomena, and relationships, by delimiting and asserting control over a geographic area'.[23] Doboš further explains that in an African context, 'political territoriality as a strategy is in many cases pursued by non-state actors challenging the

18 Caroline Varin, *Boko Haram and the War on Terror.* (ABC Clio: Praeger, 2016).
19 Jordan, David. *Understanding Modern Warfare.* (Cambridge: Cambridge University Press, 2008).
20 Varin *Boko Haram and the War on Terror*, 117.
21 Bruce Hoffman, *Inside Terrorism.* (New York: Columbia University Press, 2006).
22 Colleen Bell and Brad Evans. 'Terrorism to Insurgency: Mapping the Post-Intervention Security Terrain'. *Journal of Intervention and Statebuilding* 4 (2010): 371-390.
23 Robert Sack, Human Territoriality: Its Theory and History (Cambridge: CUP 1986), 31.

institution of state'.[24] This includes both the deliberate stated objective of controlling territory, and the ability to perform some of the state's functions within that defined space. The latter assumes a willingness to govern, not just occupy land. A crucial difference to be made between territorial actors and non-territorial actors is the ability to finance their operations through taxation, rather than looting, a return to Mancur Olson's original thesis that distinguishes between a 'sedentary' and a 'roving' bandit in his analysis of the state.[25] Non-territorial actors are dependent on foreign support criminal activities such as banditry or trafficking, or on looting the population. Territorial violent non-state actors take a more long-term strategic approach in that they engage with the population, encouraging their prosperity in order to tax them and finance their operations. As a result, they often put in place local institutions that guarantee security, offer services such as basic healthcare and education, and install bureaucratic infrastructures to tax the people living under their rule.

Warlords, Chaos and Non-State Territory

Repeatedly in the African continent, huge parcels of land have been carved up by individuals claiming the right and the ability to govern in lieu of the state. Referred to as 'warlords', these VNSAs usually emerge in an environment in which there is already a high level of violence, such as civil wars. They are mainly charismatic local leaders who, by establishing their own private army and exploiting indigenous resources, have claimed a territory within the state – often linked to their ethnic affiliation – which they claim to rule. Warlords benefit from a situation of chaos and a breakdown of public authority and impose their own rule through coercion and patronage. They have a vested interest in the security and stability of their own territory[26] and can provide enclaves of peace and governance within a conflict zone. Rather than seeking to undermine the state through acts of violence, warlords prefer the maintenance of a *status quo* of chaos and violence beyond their territory, allowing them to derive their economic

24 Doboš, 'Shapeshifter of Somalia: Evolution of the Political Territoriality of Al-Shabaab'.
25 Mancur Olson, 'Dictatorship, Democracy, and Development'. *The American Political Science Review* 87, no. 3 (1993): 567-76.
26 Kimberly Marten. 'Warlordism in Comparative Perspective'. *International Security* 31, no. 3 (2006): 41-73.

prosperity and maintain local legitimacy.[27] By all definitions, geographic space remains a central criterion in the definition and *raison d'être* of warlords.[28] However, a key criteria that distinguished them from other territorially minded non-state actors, according to Laura Freeman, is that warlords do not seek state power and do not threaten the existence of the state or the governing polity.[29]

In their volume on Violent Non-State Actors in Africa, Varin and Abubakar examine four cases that illustrate the territorial ambitions of warlords in Africa. These include the warlords in Uganda, Sierra Leone and Somalia, and the politics of piracy in East Africa, which resemble to a certain extent the territorial dynamics of warlord politics. Joseph Kony is the quintessential example of warlordism in Africa. He emerged in northern Uganda in the 1980s and at his apogee built an army of 100,000 fighters operating over an area that included parts of Uganda, Sudan, Central African Republic (CAR) and Democratic Republic of Congo (DRC).[30] According to Jo-Ansie van Wyk, Kony's objectives were focused beyond the control of his territorial enclave to include his ethnic group and cleanse the Acholi of 'evil forces'.[31] Kony's legitimacy is anchored in his messianic mission aimed at the Acholi, rather than a political overthrow of the Museveni government; but in a country that offers multiple ungoverned spaces, Kony also provided socio-economic security to the people living in his territory.[32] Kony and his Lord's Resistance Army have been exceptionally durable, adaptable and violent and despite military and international onslaughts against the warlord, Kony has remained elusive and relatively powerful within the borderlands between CAR, DRC, Sudan and Uganda.

Until he became President in Liberia (and arguably afterwards), Charles Taylor was an infamous warlord who caused a long period of anarchy in Liberia and neighbouring Sierra Leone and Ivory Coast. Along

27 Ulrich Schneckener, 'Dealing with Armed Non-State Actors in Peace- and State-Building, Types and Strategies'. *Transnational Terrorism, Organized Crime and Peace-Building.* (Palgrave Macmillan, London: 2010).

28 Laura Freeman, 'The African warlord revisited', *Small Wars & Insurgencies,* 26:5 (2015): 790-810.

29 Antonio Giustozzi, *Respectable Warlords? The Politics of State-Building in Post-Taleban Afghanistan.* Crisis States Research Centre Working Paper no. 33. (London: Crisis States Research Centre, LSE, 2003).

30 Paul Zwier, Principles negotiation and mediation in the international arena. Talking with evil. (Cambridge University Press 2013), 306.

31 Jo-Ansie van Wyk, 'Joseph Kony and the Lord's Resitance Army' in *Violent Non-State Actors in Africa* eds. Varin and Abubakar (Palgrave 2017), 229.

32 Wyk, 'Joseph Kony and the Lord's Resistance Army' in *Violent Non-State Actors in Africa* eds. Varin and Abubakar (Palgrave 2017), 242.

with other Liberian warlords, such as Prince Yomi Johnson, Taylor opposed the regime of Samuel Doe; he progressively occupied and controlled territory in resource-rich areas of Liberia, benefitted from foreign support, and instigated rebel insurgencies nationally and in neighbouring countries.[33] Unlike Kony however, Taylor had a clear goal to take power, and as such Freeman does not include the Liberian in her categorisation of warlords in Africa.[34] Tar and Wapmuk on the other hand point to Taylor's control over 'his own non-state entity beyond Liberia to Sierra Leone and Cote d'Ivoire', his lack of formal international recognition, and indiscriminate violence to characterise the Liberian as a typical warlord, one who is not characterised by his political ambitions but by his control over non-state territory and extreme levels of aggression against the people within his grasp.[35]

A third example of warlord politics defined around the concept of non-state territory is developed in Tar and Mustapha's treatment of Somalia.[36] While Al-Shabaab is an internationally designated terrorist organisation, and often the focus of study when examining the country, the politics of Somalia are generally directed by clans whose territorial control over parts of the country are evidence of warlordism. Following the collapse of Siad Barre's regime in 1991, warlords 'drummed up support among their clans, secured heavy weapons and began to unleash violence against what was left of the state'.[37] It was the utter disintegration of social order and justice that led to the emergence and popular embrace of the Islamic Court Union (ICU) and their military wing Al-Shabaab, which offered an alternative to the chaos engineered by the warlords.

The territorial link between warlords and piracy is clear in Clayton Allen's examination of pirates in Somalia.[38] He explains the geographical opportunity of Somalia's location as a key port area with considerable passage of goods in the narrow straight of the Gulf of Aiden. By controlling

---

33 Usman Tar and Wapmuk, 'The Revolutionary United Front, Liberian Warlords and Civil War in Sierra Leone' in *Violent Non-State Actors in Africa* eds. Varin and Abubakar (Palgrave 2017), 262.
34 Freeman, 'The African warlord revisited'.
35 ar and Wapmuk, 'The Revolutionary United Front, Liberian Warlords and Civil War in Sierra Leone', 268.
36 Usman Tar and Mala Mustapha 'Al-Shabaab: State Collapse, Warlords and Islamist Insurgency in Somalia' in *Violent Non-State Actors in Africa* eds. Varin and Abubakar (Palgrave 2017), 277.
37 Tar and Mustapha 'Al-Shabaab: State Collapse, Warlords and Islamist Insurgency in Somalia', 280.
38 Allen D Clayton, 'Pirates in West Africa and Somalia', Somalia' in *Violent Non-State Actors in Africa* eds. Varin and Abubakar (Palgrave 2017), 305.

adjacent territory, warlords were able to exploit their access to the sea through sponsored piracy, knitting strong clientelistic relationships with bands of pirates in the area. This provided an additional source of income by treating maritime commerce as a 'local resource' and a line of livelihood in a country racked by civil war.[39]

The space for warlords continues to thrive in Africa, with this VNSA emerging in Libya, Mali, Sudan and Central African Republic, to name a few. The weakening and disintegration of the state has created opportunities for charismatic and 'religious' leaders to carve out parcels of territorial influence. The best way to remain in control of that land, is to stay removed from the politics of the centre, a lesson than many have yet to learn.

Rebels: Holding Territory as a Means to an End?

Traditionally, rebels were armed civilians who sought to overthrow the government. They are often perceived or portrayed as liberators who fight against an unjust system, a colonial power of a corrupt or illegitimate government with the objective of freeing a territory from oppression.[40] Rebels generally have an ideological leaning and seek popular and sometimes international legitimacy as they ultimately aim to replace the government.[41] Despite claims of popular support, rebels frequently benefit from foreign military and financial support, enabling them to operate freely without concern for the well-being of the local population in the rural areas from which they govern. Holding territory is often central as both a means and an end for rebel groups.[42] De la Caille and Sánchez-Cuenca explain that the seizure of territory changes the nature of violent non-state actors in the following ways: first, territory is an effective bargaining tool against the state, enabling rebels to hold ground when trying to undermine or negotiate with the government. Second, it facilitates recruitment and power over the population settled within the borders controlled by the rebels. This

---

39 Clayton, 'Pirates in West Africa and Somalia', Somalia'.
40 Stathis Kalyvas, Kocher, Matthew Adam, 'How 'Free' is Free Riding in Civil Wars? Violence, Insurgency, and the Collective Action Problem'. *World Politics* 59 (1) (2007): 177–216.
41 Anthony Vinci, *Warlords in the International Order: A neorealist approach.* (London School of Economics, 2007), 13; Isabelle Duyvesteyn, 'Rebels & Legitimacy; An Introduction', *Small Wars & Insurgencies,* 28:4-5 (2017), 669-685.
42 Calle, Luis de la, and Ignacio Sánchez-Cuenca. 'Rebels without a Territory: An Analysis of Nonterritorial Conflicts in the World, 1970–1997'. *Journal of Conflict Resolution* 56, no. 4 (August 2012): 580–603.

enables the group to grow in size and in strength as it can draw resources from the land and its inhabitants. Third, the authors correlate the lethality of a group to their ability to control territory.[43] However, the rebel's ability to challenge the state's sovereignty is inversely proportional to the state's capacity to maintain a monopoly over the use of violence.

The centrality of territorial conquest and the role it plays in the stagnation or victories of rebels are explored in case-studies on rebel movements in CAR and Ethiopia.

In CAR, flares of rebel violence fighting for territory and political gain can be traced back to independence in the 1960s.[44] Isaacs-Martin points to the emergence of rebel groups 'vying for political power and resource access' and the political manipulations of successive national leaders who seek favour among their ethno-religious groups as the main variables behind the ongoing violence. The large reserves of natural resources such as timber and minerals have been exploited freely by competing rebel groups as a means to a political end, with the government unable to or uninterested in controlling these areas. Since 2014, four factions of ex-Seleka leaders have controlled 60 to 80 per cent of CAR's territory outside Bangui, evidence of the state's lack of capacity and of the opportunity this affords the rebel groups.[45] Unlike other rebel groups however, these militias 'have shown little focus or interest on national or regional issues',[46] and it is only through the personal ambitions of individuals, particularly that of Michel Djotodia, that they temporarily emerged out of their established zones in 2013 to occupy territory and challenge the power of the capital of CAR. Unable to control his fighters however, Djotodia soon fell from favour, and ensuing presidents have lacked the resources, despite foreign support, and arguably the resolve to negotiate a convincing peace deal with the many rebel factions around the country.

The situation is very dissimilar in Ethiopia, where 'rebel movements are currently mobilizing resources and populations in the process of a protracted armed struggle for legitimacy, political power, autonomy and

43 Calle, Luis de la, and Ignacio Sánchez-Cuenca. 'Rebels without a Territory: An Analysis of Nonterritorial Conflicts in the World, 1970–1997'.

44 Wendy Isaacs-Martin, 'The Seleka and Anti-Balaka Rebel Movements in the Central African Republic' in *Violent Non-State Actors in Africa* eds. Varin and Abubakar (Palgrave 2017), 133.

45 Council on Foreign Relations, 'Violence in the Central African Republic | Global Conflict Tracker' *Global Conflict Tracker*. (2019). https://www.cfr.org/interactive/global-conflict-tracker/conflict/violence-central-african-republic.

46 Isaacs-Martin, 'The Seleka and anti-Balaka Rebel Movements in the Central African Republic', 150.

secession'.[47] Here, territory is the ends, not the means, with rebel groups aspiring for complete separation from the central state. A select list of rebel movements in Ethiopia since 1961 shows nine rebel movements motivated by secession from the state, three have ambitions to achieve autonomy from the centre, and two, the Tigray People's Liberation Front (TPLF) and the Ethiopian People's Patriotic Front have fought for regime change.[48] The current government, however, is also evidence of rebel groups successfully overthrowing and replacing a military regime. During the civil war, the Ethiopian People's Revolutionary Democratic Front (EPRDF) was a coalition of rebel groups opposed to the government of Mengistu Halle Mariam. After 'liberating' the northern areas, the EPRDF both sought to govern these free zones and forcibly recruit the population into the 'people's struggle', giving them experience in administering 'a mini-state before it seized power and took control over all' of the country.[49] It was the successful amalgamation of regionally and ethnically based movements from around Ethiopia into one national political party that led to the military and political success of the EPRDF,[50] a stark contrast to the split factions of rebel groups and warlords in CAR. Nonetheless, regional fervour for territory, autonomy and self-determination continues to dominate the political scene in Ethiopia. Berouk Mesfin argues that this is likely to continue, as the ruling party has failed to rule in an inclusive manner, maintaining the impression that it is just 'another heavy-handed ethnocratic dictatorship', responding violently to any challenges to its power and refusing to negotiate meaningfully with existing rebel movements.[51] He concludes that a repeat of the civil war and change of power in 1991 is inevitable in the future of the country.

Before moving to an examination of the territorial ambitions of terrorists in twenty-first century Africa, an interesting case worth examining is the 'Islamist' Allied Democratic Forces (ADF) in DRC, which blurs the lines between rebels, terrorists and warlords. Initially a rebel movement with aspirations to overthrow the Ugandan government and replace it with an Islamic regime, the ADF has metamorphosed into a largely apolitical group that prioritises legal and illegal commercial activities in rural parts of

47 Berouk Mesfin, 'Rebel movements in Ethiopia' in *Violent Non-State Actors in Africa* eds. Varin and Abubakar (Palgrave 2017), 163.
48 Berouk Mesfin, 'Rebel movements in Ethiopia', 168.
49 Terrence Lyons 'From victorious rebels to strong authoritarian parties: prospects for post-war democratization', *Democratization*, 23:6 (2016), 1026-1041.
50 Lyons 'From victorious rebels to strong authoritarian parties: prospects for post-war democratization'.
51 Berouk Mesfin, 'Rebel movements in Ethiopia', 184.

eastern DRC.[52] The ADF occupied territory in the Beni area of the country, where there is little effective security presence and relatively no interest from the Congolese or Ugandan states. Jesper Cullen argues that the group has 'no notable operational capability outside of Beni',[53] partly because of the effective deployments of Ugandan security forces in the border region. The Ugandan government has sought to portray the ADF as part of a wider network of Islamist militant groups in East Africa however, labelling the group as a 'terrorist organisation' – although this may be a ploy to attract foreign financial support, as Cullen's research does not show any relationship between the ADF and other Islamist terrorist groups. The change in the group's political, ideological and military inclinations appears to be the result of internal divisions rather than external factors, and only the holding of territory remains constant. This suggests that the nature of the violent non-state actor, whether it is a terrorist, rebel group or warlord is vulnerable to change, not depending just on its ability to control territory and its political objectives, but principally on the strength and ideas of its frequently fickle leadership.

Territory as Sovereignty: The End-Goal for Today's Terrorists

As mentioned above, terrorists have been described as would-be rebels without territory,[54] non-state political agents whose use of or threat of violence undermines the credibility of the state to enforce security,[55] and violent non-state actors with largely political, religious and ideological goals who use violence to create fear among the population.[56] Only in de la Caille and Sanchez' definition is the lack of territory a key component differentiating terrorists from other violent non-state actors. In other definitions, territory is barely mentioned, unless to say that terrorists 'generally do not attempt to seize hold of territory'[57] but they prefer to

52 Jesper Cullen, ''Islamist' Rebels in DRC: the Allied Democratic Forces' in *Violent Non-State Actors in Africa* eds. Varin and Abubakar (Palgrave 2017), 109.
53 Cullen, ''Islamist' Rebels in DRC: the Allied Democratic Forces'.
54 Calle, Luis de la, and Ignacio Sánchez-Cuenca. 'Rebels without a Territory: An Analysis of Nonterritorial Conflicts in the World, 1970–1997'.
55 Yonah Alexander, *International Terrorism: National, Regional and Global Perspectives.* (New York: Praeger, 1976); Hoffman, Bruce. *Inside Terrorism* (New York: Columbia University Press, 2006).
56 Stephen Sloan, *Terrorism: The Present Threat in Context.* (Oxford: Berg Publishers. 2006).
57 Hoffman, Bruce. *Inside Terrorism* (New York: Columbia University Press, 2006), 35.

operate in urban settings, as 'acts of terror (in) urban centres are most susceptible to this form of political violence because of the likelihood of greater impact and visibility afforded by cities'. This helps to undermine the ruling regime in the eyes of its citizens. This chapter argues that terrorists in Africa in the twenty-first century have deliberately sought to occupy and govern territory as the end goal to their use of violence. This has been a repeated trend across the continent, starting with Al-Shabaab who controlled huge parts of Somalia in the early 2000s, followed by Al-Qaeda in the Islamic Maghreb's (AQIM) dominance in Mali, which was a turning point and inspiration for other aspiring Islamist terrorist groups, including Boko Haram in Nigeria and the Islamic State of Iraq and Somalia (ISIS) in Libya. The shift in aspirations has made it ever more difficult to distinguish between violent non-state actors, as designated terrorists resemble ever more the rebel groups and warlords of Africa.

Established definitions and understanding of terrorists were torn apart when, in 2012, the Islamist jihadi group AQIM took over the governance of Mali. Since its establishment in the early 2000s, AQIM's goals and objectives in North Africa and Northern Mali specifically were mainly financial, kidnapping western citizens and engaging in drug and human trafficking.[58] Jude Cocodia explains AQIM's territorial ambitions as the consequence of a 'sophisticated reading of the local context and seamless integration into the lifestyle of northern Mali'.[59] Plainly speaking, they took advantage of a weakening regime, the defeat of the Malian army by the separatist rebels, the National Movement for the Liberation of Azawad (MNLA), and their existing local alliances to push out competing factions and brutally take over the country. Arguably their territorial ambitions had a pecuniary rather than ideological motive: Northern Mali had long been the gateway for the drug trade, particularly Marijuana and Cocaine to Africa and Europe and enjoyed deep trafficking networks across the region.[60] Nonetheless, between April 2012 and January 2013, AQIM occupied and governed over northern Mali, a territory the size of France, enforcing their violent interpretation of sharia law. During their rule, AQIM and other Islamist groups attacked civilians, destroyed churches and sacred mausoleums, persecuted Christians, implementing brutal punishments, including public floggings, amputations and death for crimes such as adultery and alcohol

58 Jude Cocodia, 'Nationalist Sentiment, Terrorist Incursions and the Survival of the Malian State', in *Violent Non-State Actors in Africa* eds. Varin and Abubakar (Palgrave 2017), 49.
59 Cocodia, 'Nationalist Sentiment, Terrorist Incursions and the Survival of the Malian State'.
60 Varin, *Boko Haram and the War on Terror*, 126.

consumption.[61] They also invited and trained like-minded terrorists from across the Middle East and Africa, showing that a non-state actor, in this case with means and ambition, an Islamist terrorist group, was capable of controlling and governing territory. This lasted until the international community, led by the French, finally decided to intervene.[62] However this had already set the tone for Boko Haram, ISIS and Al-Shabaab, inspired by the territorial successes of AQIM, to operate more aggressively and ambitiously in Africa.

In 2013, territory became an end-goal for the Islamist group Boko Haram, and by January 2015, they were in control of roughly 11,780 square miles, approximately the size of Belgium or Maryland, including over 70 per cent of Borno state. Discourse around a modern Islamic Caliphate resurfaced with Boko Haram leader Abu Shekau declaring his rule over an 'Islamic State. Jacob Zenn argues that the territorial expansion of the terrorist group occurred as direct result of its ties to AQIM and the transfer of knowledge that occurred in northern Mali.[63] Nonetheless, Boko Haram's occupation of territory in Nigeria would not have been possible without the military and political disintegration of the state in northern Nigeria at the time. Boko Haram, like AQIM, was opportunistic rather than strategic in its conquest of territory. Unlike warlords in other areas of Africa, the Islamist terrorist group failed to survive as an occupying force beyond a few months and were kicked out of their safe zones by a combination of internal military forces and foreign support. They quickly absconded and resurfaced in a different form, remaining in remote areas and using terrorist tactics beyond the reach and interest of the state.

The territorial ambitions of ISIS has been well documented in recent literature. According to Larissa Jaeger, 'the efforts of Islamic State in Libya are part of the organization's greater political strategy, which includes establishing a theocratic state-Caliphate, based on strict implementation of Islamic Shari'a law throughout the Muslim World'.[64] In Libya, the territory 'held' by Islamic State has fluctuated, but at its peak included 150 miles of coastal territory, including the city of Sirte, between 2015 and 2016.[65]

61 FIDH-AMDH, 'War Crimes in Northern Mali' (2012).
62 Varin, *Boko Haram and the War on Terror*, 126.
63 Jacob Zenn, 'Boko Haram's Conquest for the Caliphate: How Al Qaeda Helped Islamic State Acquire Territory', *Studies in Conflict & Terrorism* (2018).
64 Larissa Jaeger, 'Islamic State in Libya, in *Violent Non-State Actors in Africa* eds. Varin and Abubakar (Palgrave 2017), 49.
65 Christopher Livesay and Alessandro Pavone, 'Libya's Political Instability Makes Room for ISIS to Regroup'. PBS Newshour (2019).

Giorgio Cafiero explains that 'ISIS's chief goals were disrupting Libya's hydrocarbons exports, consolidating its military gains, and expanding its territorial control'.[66] Since being pushed out of its territories in 2016 by a US-backed anti-Islamic state coalition, ISIS in Libya has engaged in terrorist tactics, including high-profile attacks on state institutions. Like AQIM, it exploits the extensive desert to develop its smuggling business, operating more like a criminal organisation.

The occupation and governing of territory clearly remain a stated and strategic goal for Islamist terrorist groups in Africa that seek to emulate the success of AQIM in 2012. Although they have mostly been pushed back by coalitions of foreign-backed domestic forces, AQIM, ISIS and Boko Haram have proven their ability to defeat the state and claim effective sovereignty, if only for a short time. In the twenty-first century, it is clear that territoriality has become a feature of terrorist groups operating in the African continent, further blurring the lines between warlords, rebels and terrorists. This needs to be taken into account by academics and policy makers when devising a strategy to address the growing threat of terrorists and a territorial violent non-state actor in Africa. Failing to incorporate their territorial ambitions into classical counter-terrorism strategies would not only be a misreading of the situation but could have grave repercussions for the security of the territory.

Conclusion

The difference between ISIS, Boko Haram, AQIM and other types of terrorist groups is that the Islamist terrorists have clearly stated their intention to usurp sovereignty and redraw existing frontiers around a new Islamic Caliphate. Not only have they voiced this goal, they have also acted on it. Since 2012, ISIS, Boko Haram and AQIM have each replaced the government and created a mini-state, which they have governed – violently but nonetheless – according to their ideals. Returning to Sack's definition of territoriality, it is clear that terrorist groups active in Africa are attempting, and succeeding, to 'influence, affect, or control objects, people, and relationships by delimiting and asserting control over a geographic

66 Giorgio Cafiero. 'Is ISIS The Real Winner Of Hifter'S Tripoli Offensive?' (Middle East Institute: 2019) https://www.mei.edu/publications/isis-real-winner-hifters-tripoli-offensive.

area'.[67] This is not unlike the behaviours of rebel groups and warlords, who share the political space in many African countries.

ISIS, Boko Haram and AQIM have been opportunistic in their quest for territory, which has changed the strategic game for other terrorist groups, as they can now realistically aspire to challenging the state and governing, like warlords, over their fiefdom. This convergence over territory, a criterion that previously distinguished terrorists from rebels and warlords in Africa, appears to make it harder to create clear categories for violent non-state actors. And yet, the political nature of labelling groups continues to supersede academic discussions, as evidenced in the case of the ADF and in 2013 when Goodluck Jonathan sought to paint Boko Haram with the same brush as Al-Qaeda in order to appeal for foreign support. Nonetheless, it may be time to find a new criterion to distinguish VNSAs in Africa, other than the possession, governance or ambition to hold territory. While they converge over their quest for territory, the overall political ambitions and methods continue to distinguish warlords, who remain fairly static, rebels that seek to replace the state and terrorists whose ambitions on the African continent transcend borders and politics. Effective counter-terrorist and counter-insurgency strategies will need to begin with a precise understanding of the nature, ambitions and methods of the enemy.

References

Bell, C. and B. Evans, "Terrorism to Insurgency: Mapping the Post-Intervention Security Terrain'. *Journal of Intervention and Statebuilding*, 4, 371-390, 2010.

Cafiero, G. 'Is ISIS The Real Winner of Hifter's Tripoli Offensive?', *Middle East Institute*, 2019.

Charter of the Organization of African Unity, 1963.

Council on Foreign Relations, 'Violence in the Central African Republic | Global Conflict Tracker', *Global Conflict Tracker*, 2019.

Crenshaw, M. 'The Causes of Terrorism'. *Comparative Politics* 13, no. 4: 379-99, 1981.

David, D. *Understanding Modern Warfare*, Cambridge: Cambridge University Press, 2008.

Davis, D. 'Non-State Armed Actors, New Imagined Communities, and Shifting Patterns of Sovereignty and Insecurity in the Modern World', *Contemporary Security Policy*, 30:2, 221-245, 2009.

---

67 Sack, *Human Territoriality: Its Theory and History*.

De la Calle, L. and I. Sánchez-Cuenca. 'Rebels without a Territory: An Analysis of Nonterritorial Conflicts in the World, 1970–1997'. *Journal of Conflict Resolution* 56, no. 4: 580–603, August 2012.

Doboš, B. 'Shapeshifter of Somalia: Evolution of the Political Territoriality of Al-Shabaab', *Small Wars & Insurgencies*, 27:5: 937-957, 2016.

Dugan, L., G. LaFree, L. De La Calle, I. Sánchez-Cuenca, V. Asal, M. G. Findley, and J. K. Young. 'Killing Civilians or Holding Territory? How to Think about Terrorism'. *International Studies Review*, 14, no. 3: 475-97, 2012.

Duyvesteyn, I. 'Rebels & Legitimacy; An Introduction', *Small Wars & Insurgencies*, 28:4-5 669-685, 2017.

Englebert, P. *State Legitimacy and Development in Africa*, Lynne Rienner, 2002.

FIDH-AMDH, 'War Crimes in Northern Mali' (2012).

Freeman, L. 'The African warlord revisited', *Small Wars & Insurgencies*, 26:5, 2015.

Giustozzi, A. *Respectable Warlords? The Politics of State-Building in Post-Taleban Afghanistan*. Crisis States Research Centre Working Paper no. 33. London: Crisis States Research Centre, LSE, 2003.

Hoffman, B. *Inside Terrorism*, New York: Columbia University Press, 2006.

Jackson, R. and C. Rosberg. 'Why Africa's Weak States Persist: The Empirical and the Juridical in Statehood' *World Politics*, Vol. 35, No. 1, 1982.

Kalyvas, S. and M. Kocher. 'How 'Free' is Free Riding in Civil Wars? Violence, Insurgency, and the Collective Action Problem'. *World Politics* 59 (1): 177–216. 2007.

Livesay, C. and A. Pavone. 'Libya's Political Instability Makes Room for ISIS to Regroup'. PBS Newshour. 2019.

Lyons, T. 'From victorious rebels to strong authoritarian parties: prospects for post-war democratization', *Democratization*, 23:6, 1026-1041, 2016.

Marten, K. 'Warlordism in Comparative Perspective'. *International Security* 31, no. 3: 41-73, 2006.

Olson, M. 'Dictatorship, Democracy, and Development'. *The American Political Science* Review 87, no, 3: 567-76. 1993.

Oppenheim, L. *International Law* Vol – 1. 8'11 cd. 451-52, 1955.

Sack, R. *Human Territoriality: Its Theory and History*, Cambridge: CUP 1986.

Schneckener, U. 'Dealing with Armed Non-State Actors in Peace- and State-Building, Types and Strategies'. *Transnational Terrorism, Organized Crime and Peace-Building*. Palgrave Macmillan, London: 2010.

Shaw, M. 'Territory in International Law'. *Netherlands Yearbook of International Law*, 13, 61-91, 1982.

Sloan, S. *Terrorism: The Present Threat in Context*. Oxford: Berg Publishers. 2006.

Tilly, C. *Capital, Coercion, and European States, AD 990-1992*, Cambridge: Basil Blackwell, 1990.

Varin C. and A. Dauda. *Violent Non-State Actors in Africa: Terrorists, Rebels and Warlords*, Palgrave Macmillan, 2017.

Varin, C. *Boko Haram and the War on Terror*. ABC Clio: Praeger, 2016.

Vinci, A. *Warlords in the International Order: A neorealist approach*, London School of Economics, 2007.

Weber, M. *The Protestant Ethic and the Spirit of Capitalism,* New York: Scribner's Sons, 1958.

Yonah Alexander, Y. *International Terrorism: National, Regional and Global Perspectives.* New York: Praeger, 1976.

Zenn, J. 'Boko Haram's Conquest for the Caliphate: How Al Qaeda Helped Islamic State Acquire Territory', *Studies in Conflict & Terrorism,* 2018.

Zwier, P. *Principles negotiation and mediation in the international arena. Talking with evil.* Cambridge University Press 2013.

2

WHO GETS TO FIGHT?

The De-Legitimation of Mercenaries and the Search for the Ideal Polity

Helene Olsen

Introduction

Mercenaries, as non-state actors, have been used extensively throughout history. Yet, as actors of war, they have consistently been perceived in a negative light, as bad people and bad for society – actors who should be far removed from the battlefield and who cannot legitimately participate in war. Sarah Percy explains that there has been a norm against mercenaries alongside their extensive use and Janice Thomson highlights that mercenaries were actively de-legitimised when states tried to consolidate the legitimate use of violence in the late eighteenth-century.[1] Mercenaries have been labelled the 'scourge of western Europe', and Machiavelli famously declared that mercenaries are 'useless and dangerous' individuals who cannot 'be trusted'.[2] These criticisms point towards a perception of mercenaries as illegitimate participants of war. However, this chapter argues that mercenaries are not inherently illegitimate. Rather, they are constructed as such through distinct discursive practices that facilitate the

1 Sarah Percy, *Mercenaries: The History of a Norm in International Relations* (Oxford: Oxford University Press, 2007), 1; Janice E. Thomson, *Mercenaries, Pirates, and Sovereigns: State-Building and Extraterritorial Violence in Early Modern Europe.* (Princeton, N.J.: Princeton University Press, 1994), 88.

2 Kenneth Fowler, *Medieval Mercenaries*, Vol 1: The Great Companies. (Oxford: Blackwell Publishers, 2001), ix; Niccolò Machiavelli, *The Prince*, ed. Robert M. Adams, 2nd ed. (New York & London: Norton & Company, 1992), bk. XII, 34.

conceptual understanding of mercenaries as illegitimate fighters. These discursive practices are speech acts – discursive acts such as 'making statements, giving commands, asking questions, making promises' – which come in different forms and are not confined to text.[3] Therefore, in this chapter the focus will be on several different forms of utterings, both written and spoken. Speech acts are essentially performative acts where words are not just 'saying something' but also 'doing something'.[4] In this way, speech acts are performances that, according to Pierre Bourdieu, can 'modify social reality' and 'contribute to its production' – speech acts can construct social reality.[5] Thus, speech acts perform and perpetuate the understanding of mercenaries as illegitimate fighters, and mercenaries are thus assigned illegitimacy through the repeated usage of a specific set of identifiable speech act 'moral objections' expressed in connection with their usage. Both J.L. Austin and Pierre Bourdieu highlight that the success of the speech act is dependent on the authority of the individual doing the uttering and the circumstances in which the uttering is done.[6] In this way, the individuals who construct mercenaries as illegitimate fighters via moral objections have a certain standing within their respective societies.

In this chapter, it will further be suggested that mercenaries have been most violently objected to and de-legitimised in situations where their presence on the battlefield has threatened prevalent conceptions of the ideal polity. As such, when mercenaries have presented a threat to employers' understanding of their own polity – regardless of the type of polity – mercenaries' presences have been morally objected against and they have actively been de-legitimised. As Bourdieu emphasises, 'the constitutive power of…language, and the schemes of perception and the thought which it procures, is never clearer than in situation of crisis: these *paradoxical* and *extra-ordinary* situation call for an extra-ordinary kind of discourse'.[7] In this way, in situations where the use of mercenaries is perceived as extra-ordinary or where mercenaries themselves are deemed disruptors of the ideal polity, they have been morally objected against and de-legitimised. Thus, the implication of the de-legitimation of mercenaries is not just that they are conceptualised as illegitimate fighters; the process of de-legitimisation

3 John R. Searle, *Speech Acts: An Essay in the Philosophy of Language.* (Cambridge: Cambridge University Press, 1969), 16.

4 Searle, 22; J.L. Austin, *How to Do Things with Words.* (Cambridge, MA: Harvard University Press, 1975), 14–15, 25.

5 Pierre Bourdieu, *Language and Symbolic Power.* (Cambridge: Polity Press, 1991), 128.

6 Austin, *How to Do Things with Words*, 34; Bourdieu, *Language*, 170.

7 Bourdieu, *Language*, 128–29; emphasis original.

also provides us with valuable information about the relationship between mercenaries, their employers, and the surrounding polity. The argument will be presented as follows; firstly, it will be discussed what legitimacy means in relation to mercenaries and which types of individuals can be considered as mercenaries. Secondly, the two categories of moral objections made towards mercenaries will be presented: deontological, which focus on the characteristics of the individual mercenary, and consequentialist, which focus on the consequences employing mercenaries might have. This section seeks to explain how these moral objections have been more prevalent when the mercenary has been perceived to threaten the conceptions of the ideal polity, and how these moral objections have constructed the mercenary as an illegitimate fighter.

Constructing the Mercenary

This section will examine how the construction of mercenaries as illegitimate fighters is possible by first exploring what is meant by legitimacy and why legitimacy matters in the context of war and the use of force. Secondly, it will be discussed which actors can be considered a mercenary. This chapter focuses on legitimacy in relation to the use of force and as emanating from the polity. The use of force and the legitimacy of this force rests on the assumption that the state or a similar polity are legitimate perpetrators of violence.[8] As agents of a polity, fighters, whether regular armed forces, auxiliary forces, or mercenaries, derive their authority and legitimacy to use force from the polity that they serve. The use of force is therefore intimately linked to whichever power is directing and authorising this force. As such, fighters' perpetration of force does not possess legitimacy separated from the source of direction and authorisation. It is thus in relationship with the polity that fighters are constituted as either legitimate or illegitimate. Fighters are therefore not legitimate or illegitimate in their own right but are constructed as such in relation to the polity. In this way, the polity is the referent object in the construction of fighters as either legitimate or illegitimate. As will be evident from the following section, the moral objections made towards mercenaries are uttered with reference to the polity, for example, the consequences *for* the polity by employing mercenaries or the implications of the perceived lack of loyalty of mercenaries *towards* the polity.

8 Martha Lizabeth Phelps, 'Doppelgangers of the State: Private Security and Transferable Legitimacy', *Politics & Policy* 42, no. 6 (1 December 2014): 826.

In this chapter, mercenaries are not understood as 'lone dogs of war' or freedom fighters. Rather, the employer of the mercenary must have a certain degree of authority and legitimacy to (legitimately) prosecute war. Furthermore, this chapter broadly refers to a wide range of individuals as 'mercenaries'. This is a conscious choice made to highlight how their contemporaries have perceived these actors and delegitimised them using the moral objections discussed below, and the term is therefore not used to refer to specific characteristics of these individuals. If an individual has been labelled a mercenary by contemporaries, and delegitimised using the moral objections highlighted below, that individual will be described as a mercenary in this chapter. As such, Private Military and Security Companies (PMSCs), or variations of these companies, and their employees will also be considered mercenaries in this chapter, alongside more historical figures. In this way, while acknowledging that the 'mercenary' is considered a distinct category in modern international law,[9] this chapter aligns itself with works by Malte Riemann, Hin-Yan Liu and Christopher Kinsey, who argue that the mercenary is not a trans-historical figure with distinct universal characteristics.[10] Instead, what connects the figures discussed in this chapter is how the surrounding society has reacted to them, rather than characteristics such as foreignness or monetary motive. As such, while the appearance and manifestation of PMSCs might differ from a seventeenth-century military entrepreneur, how the surrounding society has reacted to these actors has remained remarkably consistent. Percy described this consistency as a prevalent proscriptive norm against mercenary use, and she has argued that the norm has affected the private military and security industry both in terms of the services these companies provide and how the companies view and market themselves.[11] Likewise, Elke Krahmann has highlighted how PMSCs have consciously tried to escape the mercenary label in order to seem more legitimate.[12] As such, the reaction to PMSCs and their employees are often marked by the same moral objections as more 'traditional' mercenaries and they have subsequently been constructed as

9 For example, in Article 47 of the Additional Protocol of the 1949 Geneva Conventions.

10 Malte Riemann, '"As Old as War Itself"? Historicizing the Universal Mercenary', *Journal of Global Security Studies*, 26 February 2020, 1–16; Hin-Yan Liu and Christopher Kinsey, 'Challenging the Strength of the Antimercenary Norm', *Journal of Global Security Studies* 3, no. 1 (1 January 2018): 93–110.

11 Percy, *Mercenaries*, 49, 227–31.

12 Elke Krahmann, 'From 'Mercenaries' to 'Private Security Contractors': The (Re) Construction of Armed Security Providers in International Legal Discourses', *Millennium: Journal of International Studies* Vol. 40, no. 2 (2012): 343–363.

illegitimate fighters. These modern mercenaries will therefore be treated in this chapter alongside their historical counterparts.

Moral Objections

Ultimately, moral objections demarcate the legitimate fighter from the illegitimate fighter in the context of war. The moral objections therefore act not only as a way to articulate how the illegitimate fighter is perceived as such, but they also illustrate what the legitimate fighter *should* be, and thereby *by whom* and *how* war is thought to be prosecuted properly. These normative judgements point towards what the ideal polity is thought to entail and by which actors this ideal polity should be protected. The moral objections made towards mercenaries have been most forcefully uttered in situations where these actors are thought to threaten the prevailing conceptions of the ideal polity. Below, two categories of moral objections will be examined, and examples where mercenaries have been thought to threaten the ideal polity will be highlighted. Rough division into categories and sub-categories is difficult, since the moral objections are intertwined and play into each other. Nevertheless, for clarity, the moral objections have been divided into two categories: deontological and consequentialist moral objections. The deontological moral objections typically come in the form of objections to mercenaries based on their monetary motive and foreignness and are therefore focused on *being* mercenary. The consequentialist objections typically come in five sub-categories: 1) lack of control, 2) human rights violations, 3) erosion of core socio-political values, 4) undermining the relationship between ruler and ruled, and 5) epistemic power. The sub-categories all focus on the perceived consequences of employing mercenaries and will be discussed in turn.

Deontological Moral Objections

The common, negative sentiment expressed in relation to mercenaries is that 'someone who hires his gun to the highest bidder, or less dramatically, fights predominantly for money will typically lack the motive appropriate to war'.[13] The idea is that mercenaries do not hold an appropriate and proper motive

13 C.A.J. Coady, 'Mercenary Morality', in *International Law and Armed Conflict*, by A.G.D. Bradney (Stuttgart: Franz Steiner Verlag, 1992), 63.

for participating war because their – presumed – predominant motivation is monetary gain. The sentiment is echoed in Article 47 of the Additional Protocol of the 1949 Geneva Conventions. This document defines a mercenary as someone who, among other attributes, 'is motivated to take part in hostilities essentially by the desire for private gain'.[14] This definition was adapted during the 1977 session of the Diplomatic Conference on Humanitarian Law in Geneva and is a definition that still stands today.[15] While Article 47 was the first 'successful international effort to define mercenaries legally', this legal definition of the mercenary is also a speech act that constructs the mercenary as an illegitimate fighter based on the perception that someone who participates in war *should not* be primarily motived by monetary gain.[16] Therefore, while Article 47 created a distinct and rather limited legal category, it was also a discursive construction of the mercenary as an illegitimate fighter. Articulated in a time when the use of predominantly white mercenaries in the twentieth-century independence wars in Africa was seen as a gross violation of self-determination efforts, Article 47 was an attempt to create a legal tool to prosecute mercenaries and to dissuade participation in mercenarism.[17] This speech act was therefore uttered in a situation where prominent conceptions of the ideal polity was challenged and the need to address these actors felt acute. The objection to mercenaries' monetary motive is often linked to the assumption that they should have a more proper motive for participating in war. This proper motive is closely associated with the conception that fighting in war should happen out of a sense of duty.[18] The ideal is that participating in war should happen based on an allegiance to a community rather than the aim for monetary reward. Fighting for one's polity is not only considered a cornerstone of the 'social contract', but being motivated by 'patriotic values' while making the ultimate communal sacrifice (i.e., dying) is seen as morally elevated.[19] As such, the deontological moral objections against mercenaries are not just founded on the perception that mercenaries possess an impure

14 International Committee of the Red Cross, 'Protocol Additional to the Geneva Conventions of 12 August 1949, and Relating to the Protection of Victims of International Armed Conflicts (Protocol I)' (Article 47, 8 June 1977).
15 H.C. Burmester, 'The Recruitment and Use of Mercenaries in Armed Conflicts', *The American Journal of International Law* Vol. 72, no. 1 (1978): 39.
16 Jeffrey S. Morton and Presley Jones, 'The Legal Status of Mercenaries', *Politics & Policy* Vol. 30, no. 4 (2002): 624.
17 Krahmann, 'From "Mercenaries" to "Private Security Contractors"', 348–50.
18 James Pattison, *The Morality of Private War: The Challenges of Private Military and Security Companies.* (Oxford: Oxford University Press, 2014), 38.
19 Andrew Alexandra, 'Private Military and Security Companies and the Liberal Conception of Violence', *Criminal Justice Ethics* 31, no. 3 (2012): 162.

monetary motive for participating in war, but also on the notion that 'private soldiers cannot be compelled to make the ultimate sacrifice when necessary'.[20] Because mercenaries cannot be expected to be driven by a sense of duty or a personal stake in the violence perpetrated, they do not belong on the battlefield.

The deontological moral objections made towards mercenaries are not just based on moral concerns, but also on a practical fear that without the patriotic motivation for participating in war, mercenaries are more likely to be disloyal to their employers. This sentiment is famously expressed by Machiavelli in *The Prince,* when he declares that mercenaries have 'no fear of God, nor loyalty to men'.[21] This opinion was shared by the British MP Colonel Isaac Barré, who, during a debate in the House of Commons in 1776 on the use of German mercenaries in America, pointed out the 'great danger of introducing such numbers of foreigners into the Kingdom'.[22] These two speech acts are examples of situations where mercenaries have been constructed as illegitimate fighters based on their foreignness and, thus, their (perceived) disloyalty.

In this way, the deontological moral objections made towards mercenaries are founded on the perception that mercenaries have an impure monetary motive for participating in war, and that their allegiance to their own monetary motive will count higher than a commitment to their employers. The fear is thus not only founded in a belief that mercenaries will be disloyal, as Machiavelli claimed, but also in the perception that the lack of attachment to their employers will make the mercenaries behave outside the proper conduct of war. The deontological moral objections made towards mercenaries are thus focused on the inner life of these actors – their motivations, values, ambitions, and intentions. These moral objections are both focused on the (presumed) monetary motive of mercenaries and the inherent foreignness of these actors.

Consequentialist Moral Objections

1. Lack of Control

The first of the consequentialist moral objections is focused on a perceived lack of control. Control over the use of force is central to any polity. In the modern conception, Max Weber famously argued that a successful state is

---

20 Pattison, *Morality of Private War*, 8.

21 Machiavelli, *The Prince*, bk. XII, p. 34.

22 *The Parliamentary History of England, from the Earliest Period to the Year 1803*, vol. 18 (1774-1777) (London: Printed by T.C. Hansard, 1813), 1181.

one that 'claims the monopoly of the legitimate use of physical force'.[23] As such, a successful state, according to Weber, is one that has control over the use of force within its territory; as actors outside the formal structures of the polity, there is no room for mercenaries in this conception. The desire to control mercenaries has a long history; we need only look at the Hundred Years' War, when periods of truces left unemployed mercenaries 'ravaging the countryside' with mercenaries consequently being called 'the scourge of western Europe', a menace that needed to be controlled.[24] The opposition, and moral objection, to these individuals was not necessarily directed at them working for money, which was a common feature of twelfth century warfare, as feudalism declined. Rather, mercenaries were objected to because they were working outside the control of the polity.[25] This made mercenaries and their conduct illegitimate.

When mercenaries in peacetime during the Hundred Years' War started forming Great Companies of organised groups of mercenaries, these started to spill over to other parts of Europe. Especially in Italy, where they became known as *condottieri*, and where several attempts were made to control them by both the surrounding society and their employers.[26] The *condotta* signed between mercenaries and their employer had evolved by the early fifteenth century into a complex contract issued for more than six months, and these contracts typically included clauses meant to prevent the *condottiere* from attacking their former employer for a period of six months, in return for payments.[27] These attempts to control the early modern mercenaries led Machiavelli to argue that mercenaries were 'undisciplined and untrustworthy', echoing a contemporary fear of a lack of control with these individuals.[28] This speech act from Machiavelli was uttered in a time when Italian humanists were in disagreement over how the ideal polity should be arranged, and who *should* fight to protect the polity if not the mercenary.[29] A consensus developed preferring the 'citizen solider' to the illegitimate mercenary, although who this citizen

23 Max Weber, *From Max Weber: Essays in Sociology.* (London & New York: Routledge, 2013), 78.
24 Percy, *Mercenaries*, 78; Fowler, *Medieval Mercenaries*, ix.
25 John France, 'Introduction', in *Mercenaries and Paid Men: The Mercenary Identity in the Middle Ages*, by John France (Leiden: Brill, 2008), 7.
26 Michael Mallet, *Mercenaries and Their Masters: Warfare in Renaissance Italy.* (London: Bodley Head, 1974), 27.
27 Mallet, 83, 86-7; C.C. Bayley, *War and Society in Renaissance Florence: The de Militia of Leonardo Bruni.* (Toronto: University of Toronto Press, 1961), 14.
28 Machiavelli, *The Prince*, 34.
29 Bayley, *War and Society*, 178.

soldier was, or should be, was a point of contestation.[30] The Renaissance scholar Petrarch and the humanist Coluccio Salutati shared the perception of mercenaries as illegitimate fighters, but were in disagreement over who then *should* fight for the polity.[31] There was resistance against 'citizen soldiers', because it was seen as 'dangerous to arm subjects', but objections against mercenaries based on the lack of control were also prevalent.[32] As such, when Machiavelli wrote *The Prince*, it was a perilous time for the Italian city-states and a time when mercenaries were seen as disruptors of the ideal polity.

In the modern conception of mercenary usage, the issues of control – or lack thereof – is closely related to questions about the regulation of the private military and security industry, as well as questions about how to make the use of mercenaries more effective and efficient, and thus more reliable. The lack of proper control and regulation is thought to have possible negative consequences on states' monopoly on violence but concerns about the possible infringement of democratic values such as accountability and transparency are also at the forefront in modern objections. To remedy these problems, various authors have suggested creating better provisions to address the 'regulatory vacuum' and negate possible negative effects of mercenaries.[33] The UN Working Group on the Use of Mercenaries – which called current regulation 'patchy and inconsistent'[34] – as well as the British Foreign and Commonwealth Office and the industry itself, have all strived to address the lack of regulation.[35] In this way, there have been official responses, both from the industry itself and from official bodies, as to the lack of regulation.

Moral objections expressed against mercenaries based on the lack of control in this area and the belief that mercenaries *should* be controlled can

30 Bayley, 181.
31 Bayley, 186–87, 192, 194.
32 Mallet, *Mercenaries and Their Masters*, 209.
33 Kevin O'Brien, 'What Should and What Should Not Be Regulated?', in *From Mercenaries to Market: The Rise and Regulation of Private Military Companies*, by Simon Chesterman and Chia Lehnardt (Oxford: Oxford University Press, 2007); Sarah Percy, 'Regulating the Private Security Industry', Adelphi Paper Vol. 384 (London: Routledge & IISS, 2006).
34 United Nations Commission on Human Rights, 'Report of the Working Group on the Use of Mercenaries as a Means of Violating Human Rights and Impeding the Exercise of the Right of Peoples to Self-Determination' (A/HRC/36/47, 20 July 2017), 17.
35 British Foreign and Commonwealth Office, 'Private Military Companies: Options for Regulations', Green Paper (London: House of Commons, 2002); Paul McGrade, 'Private Military and Security Companies', *Chatham House: International Law Summary*, 2013, 10; International Stability Operations Association, 'ISOA Code of Conduct', § Version 13.1, https://stability-operations.org/page/Code.

thus be traced back the Hundred Years' War and the use of mercenaries in Renaissance Italy. Along the same lines, concerns with the use of mercenaries due to the loss or lack of control are also expressed in connection with the modern private military and security industry.

2. Human rights violations

The second sub-category of consequentialist moral objections focus on the possible human rights violations that mercenaries might commit, as well as the overall negative impact that the use of mercenaries might have on the environment in which they operate. The overall fear expressed in this sub-category is that because mercenaries are supposedly motived primarily by monetary gains, and not by patriotic duty or a 'higher cause', they will be less constrained by the prevailing laws of war and are therefore more likely to commit wrongdoings that are outside the proper prosecution of war. Because mercenaries are 'private' actors existing outside the polity, they are supposedly exempted, both practically and morally, from many constraints that would prevent human rights violations. This sub-category is, however, not just focused on the loss of innocent lives, but also on how the behaviour of mercenaries can negatively affect their employers. In a modern setting, the argument is typically formulated in the following way: 'If soldiers misbehave they are subject to significant penalties; if employees of a PMC [Private Military Company] operating on foreign soil misbehave, they are, at worst, subject to being fired from the job. Thus, opportunistic behaviour of PMC employees is more likely to occur than among soldiers'.[36] The objection is therefore that, while the monetary motive of mercenaries makes them more likely than regular soldiers to act outside the proper conduct of war and commit human rights abuses or violate international law, they also face a lesser punishment for doing so. Ultimately, it is assumed that the monetary motive of mercenaries is the very foundation of their behaviour and that this monetary motive will lead to human rights violations because mercenaries have no regard for anything, or anyone, other than their predicted monetary reward.[37] The fear surrounding the possible wrongdoings committed by mercenaries is also closely related to control. It has thus been suggested that 'a much tighter system for vetting could reduce the likelihood of contractors violating civilians' human rights'

36 Jurgen Brauer and Hurbert van Tuyll, *Castles, Battles, and Bombs: How Economics Explains Military History.* (Chicago & London: The University of Chicago Press, 2008), 316.
37 Pattison, *Morality of Private War*, 38–39.

which would protect not only innocent civilians but also the employers that hire these actors.[38] These concerns were made explicit in the aftermath of the 2007 Nisour Square incident, when 17 Iraqi civilians were killed by Blackwater employees. The incident created what has been called a 'political firestorm,' with the Iraqi Government calling the behaviour of the Blackwater employees 'a crime'.[39] Moral objections towards mercenaries because of their effect on the surrounding environment and the negative implications for the hiring polity is also not a modern worry. In the twelfth century Gesta Stephani, a chronicle of the reign of King Stephen of England, mercenaries were described as 'profane scoundrels…shamelessly guilty of murder and pillage and various abominations, and so savagely and brutally did their reckless and unblushing presumptions rage without pity against all'.[40] The same chronicle refers also to the 'savage crowd of barbarians' who 'conspired with one mind to commit crime and outrage…[by] pillaging the goods of the poor'.[41] In the same vein, the Canons of the Third Lateran Council (1179), excommunicated mercenaries and those who hired them because of their general cruelty and indiscriminate violent behaviour.[42] In 1364, Pope Urban V, in his bull Cogit nos, further excommunicated mercenaries and accused them of being 'fired by the flames of their own greed [and] dishonestly attempting to gorge themselves on the labour of others, and for that reason rage the more cruelly against the innocent peoples'.[43] These utterings are speech acts that constructed mercenaries as illegitimate fighters. Because these individuals were motivated by private monetary gain, it was argued, they were more likely to commit human rights abuses and behave indiscriminately in a theatre of war. It is therefore possible to trace speech acts that construct mercenaries as illegitimate fighters based on their conduct in warfare to the reign of King Stephen in the twelfth-century and see the resemblance with concerns expressed over the modern usage of mercenaries.

---

38 Pattison, 8.
39 Human Rights First, 'State of Affairs: Three Years After Nisoor Square' (Human Rights First, September 2010); Sabrina Tavernise, 'U.S. Contractor Banned by Iraq Over Shootings', *The New York Times*, 18 September 2007, https://www.nytimes.com/2007/09/18/world/middleeast/18iraq.html.
40 K. R. Potter, trans., *Gesta Stephani.* (London: Nelson, 1955), 153.
41 Potter, 102.
42 Petti Joenniemi, 'Two Models of Mercenarism: Historical and Contemporary', *Instant Research on Peace and Violence* Vol. 7, no. 3/4 (1977): 189.
43 Fowler, *Medieval Mercenaries*, 119.

3. Erosion of core socio-political values

The foundation of any polity is its socio-political values. These act as firm pillars against which the actions of both those in power and the inhabitants of a polity are judged. In modern liberal democratic societies, values such as accountability, transparency, and public consent are considered core tenets. Some argue that the use of mercenaries erodes these values, with some commentators arguing that the use of mercenaries challenges the very 'meaning of a nation-state'.[44] When examining the socio-political value 'transparency', it is often argued that using mercenaries makes it difficult to obtain information about these actors and their role in warfare and that transparency is hampered as a consequence.[45] The domain in which mercenaries work is inherently secretive, which has several implications. In the United States (US), for example, commercial laws significantly reduce the possibility of obtaining detailed information about these actors, their contracts, and their activities, while information about the US military that is deemed relevant to the public is released in accordance with the Freedom of Information Act, when requested.[46] In addition, the private nature of these companies further hinders the ability to increase transparency in relation to the contract and bidding processes, meaning that it often evades democratic oversight. Thus, the procedures surrounding these actors, especially the introduction of sole bidding for governmental contracts, ultimately evades the core values of the democratic state.[47] The ability to hold individuals and larger entities accountable for their actions in the service of the state is a core value in democratic states. When military and security functions are outsourced to private actors, the lack of public oversight erodes the possibility of holding these actors accountable for their actions and increases the possible mismanagement of the contract. A fear of losing control when moving military and security functions to a 'private' actor is therefore at the heart of this objection. However, this objection is also focused on discovering and prosecuting any wrongdoings committed by mercenaries employed by the hiring polity. This desire was exemplified by the Nisour Square incident. Following the deaths of

44 Kateri Carmola, *Private Security Contractors and New Wars: Risk, Law and Ethics.* (London: Routledge, 2010), 14.

45 Deborah Avant and Lee Sigelman, 'Private Security and Democracy: Lessons from the US in Iraq', *Security Studies* Vol. 19, no. 2 (2010): 231–32.

46 Avant and Sigelman, 243–44.

47 Martha Minow, 'Outsourcing Power: How Privatizing Military Efforts Challenges Accountability, Professionalism, and Democracy', *Boston College Law Review* Vol. 46, no. 5 (5) (2005): 995.

17 Iraqi civilians, the question remained of how to hold the responsible individuals accountable. Several avenues of prosecution were impossible, including Iraqi domestic law, due to 'Order 17', which ensured that the military contractors working in Iraq would be exempted from prosecution under Iraqi law, issued by Paul Bremer in December 2006 before he left Iraq and the position as leader of the Coalition Provincial Authority.[48] However, American law and The Military Extraterritorial Jurisdiction Act (MEJA) were possibilities.[49] Passed in 2000, MEJA makes it possible for US federal courts to have jurisdiction over individuals that commit crimes while being 'employed by or accompanying the Armed Forces outside the US'.[50] However, because Blackwater was employed by the US Department of State rather than the Department of Defense, MEJA was not applicable. This left an accountability gap that was addressed in several ways, including by the passing of the 'MEJA Expansion and Enforcement Act' on 3 October 2007. This act sought to make MEJA applicable for all individuals under contract with the US in Iraq. During the deliberations for the act, Rep John Hall (D-NY) stated that he was 'proud to cosponsor and vote for the MEJA Expansion and Enforcement Act to bring these contractors under US jurisdiction if they commit criminal acts. Only by holding these contractors accountable can we actually begin to restore our standing in the world and win hearts and minds in Iraq'.[51] This statement by a US elected official was a speech act that constructed Blackwater employees and those like them as illegitimate fighters on the basis of the inability to hold these individuals accountable and thereby eroding one of the core socio-political values.

It has not only been in the modern usage of mercenaries that we can find objections against these actors based on their erosion of core socio-political values. In the late Middle Ages, fighting for a just cause under legitimate authority was the backbone of legitimate war-making.[52] As such, when Mérigot Marchès, a *routier* captain who in the 1370s and 1380s roamed southern France, was found to have acted outside legitimate authority, he was tried and sentenced to death in Paris, in 1391. The judges came to this sentence because Marchès had continued fighting after a truce

48 David H. Chen, 'Holding 'Hired Guns' Accountable: The Legal Status of Private Security Contractors in Iraq', *Boston College International & Comparative Law Review* Vol. 32 (2009): 106.
49 Chen, 106.
50 Congressional Records, 'Military Extraterritorial Jurisdiction Act of 2000', Pub. L. No. 106–523 (2000), https://www.congress.gov/bill/106th-congress/senate-bill/768/text.
51 Congressional Records, 'House Adopts MEJA Expansion and Enforcement Act of 2007' (2007), https://fas.org/irp/congress/2007_cr/h-meja.html.
52 Percy, *Mercenaries*, 72.

had been called, and Marchès was therefore deemed and constructed as an illegitimate fighter not because of the way he prosecuted war, but because he continued fighting when he no longer had legitimate authority.[53] While what can be considered core socio-political values have changed over time with different conceptions of the ideal polity, the fear that mercenaries would violate these values, and thus disrupt the ideal polity, has been a foundation of the construction of the mercenary as an illegitimate fighter since the late Middle Ages.

4. Undermine the relationship between ruler and ruled

One of the most common arguments against the use of mercenaries is that they infringe upon and have devastating consequences for the special relationship that exists between ruler and ruled, where the ruler provides protection and basic security functions for the ruled. Within this consequentialist sub-category, the perception is that the legitimate use of violence is 'the domain of the modern state, which, as a natural consequence, delegitimizes non-state providers of security'.[54] As such, the use of mercenaries is considered to be a practice that erodes states' monopoly on violence.[55] In this way, the use of mercenaries fundamentally changes the meaning of the modern state as a producer and organiser of violence. When mercenaries are used as producers of violence, the state breaks the social contract with the members of the polity.[56] Because mercenaries are not members of the polity, at least not in a formal sense, they have no stake in the social contract, and can therefore not be seen as legitimate agents of security functions.[57] Mercenaries are thus illegitimate actors that undermine the relationship between ruler and ruled by challenging the social contract. The erosion is both on a practical level, where security and military functions are no longer the product of state techniques, as well as

---

53 M. H. Keen, *The Laws of War in the Late Middle Ages.* (London: Routledge & Keegan Paul, 1965), 98–100.

54 Thomas Mandrup, 'Privatisation of Security: The Concept, Its History and Its Contemporary Application' (Copenhagen, Denmark: Royal Danish Defence College, 2012), 8.

55 Mary Kaldor, *New and Old Wars: Organized Violence in a Global Era,* Third Edition (Cambridge: Polity Press, 2012), 6; Anna Leander, 'Globalisation and the Eroding State Monopoly of Legitimate Violence', Working Papers (Copenhagen Peace Research Institute (COPRI)) (Copenhagen, Denmark: Copenhagen Peace Research Institute, 2001), 2.

56 P.W. Singer, *Corporate Warriors: The Rise of the Privatized Military Industry.* (Ithaca, N.Y.: Cornell University Press, 2008), 226.

57 Andreas Krieg, 'Towards a Normative Explanation: Understanding Western State Reliance on Contractors Using Social Contract Theory', *Global Change, Peace & Security* Vol. 25, no. 3 (1 October 2013): 350.

on a philosophical level, where the use of mercenaries challenges the very meaning of the state.

For an example of when the use of mercenaries was thought to undermine the relationship between ruler and ruled, one can look at the British use of Hessians and other German subsidiaries during the American War of Independence. After the initial fighting broke out, in 1775, it quickly became apparent that the British forces in North America needed to be supplemented with mercenaries, due to a shortage of available soldiers to fight the American rebellion.[58] In February 1776, the Prime Minister, Lord North, suggested using German mercenaries from Brunswick, Hesse-Cassel, and Hanau.[59] The debates in the Houses of Parliament that followed the suggestion of using mercenaries to fight 'our domestic quarrels', as it was termed by Lord Camden, were heated and intense.[60] Those against the move pointed towards the devastating effect that the use of mercenaries would have on the relationship between ruler and ruled. In the House of Commons, David Hartley, MP for Kingston-Upon-Hull, called the introduction of mercenaries into the conflict with the Americans 'most disgraceful,...most unjust and unnatural,' and highlighted that the suggested course of action would remove 'all possibility of reconciliation and return to our former connextion [*sic*]'.[61] Frederick Bull, MP for the City of London, declared that the mercenaries had been 'hired to subdue the sons of Englishmen and of freedom' and lamented that the measures taken against America were 'inimical...to the honor, faith, and true dignity of the British nation'.[62]

These are speech acts that construct the German mercenaries as illegitimate fighters due to the perception that their fighting in America would undermine the relationship between ruler and ruled; their introduction challenged conceptions of who *should* fight and ideas about the ideal polity. Using mercenaries against the American colonists, who were British subjects, was deeply disturbing for many members of Parliament. This objection to mercenaries came at a time of upheaval with conflicting conceptions of what constituted a state and what *should* constitute a state. As such, the mercenaries were constructed as illegitimate fighters because

58 Correlli Barnett, *Britain and Her Army: A Military, Political and Social History of the British Army, 1509 – 1970.* (London: Cassell & Co., 2000), 215; Anthony Mockler, *Mercenaries* (London: Macdonald, 1969), 107.
59 *The Parliamentary History of England,* 18 (1774-1777): 1167.
60 *The Parliamentary History of England,* 18 (1774-1777): 1222–26.
61 *The Parliamentary History of England,* 18 (1774-1777): 1170–75.
62 *The Parliamentary History of England,* 18 (1774-1777): 1185.

of the perception that the introduction of mercenaries into the mix would undermine the relationship between ruler and ruled.

5. Epistemic Power

The final sub-category of the consequentialist moral objections focuses on the fear that mercenaries as security actors in their own right can shape understandings of security discourses and the meaning of security in general. It is argued that as both producers and experts of violence, mercenaries hold a special epistemic power, a power to shape the knowledge surrounding security and violence.[63] By using mercenaries, employers are 'reproducing' mercenaries' 'authority' in the field of security, and thus allowing mercenaries to claim 'expertise and legitimacy' as security experts.[64]

It has been argued by Anna Leander and others that this practice has several negative consequences for the polity. Firstly, by using mercenaries, the role of security experts is transferred from the polity to mercenaries and polities are in this way reproducing and reaffirming these actors as security experts while foregoing their own status as security experts and providers. Polities ultimately confirm the authority of mercenaries as 'specialists of violence' as argued by Leander with reference to Harold Laswell's *Garrison State*.[65] The fear is that these actors will gain the upper-hand as experts of security and military functions in a way that might be disadvantageous to the hiring polity.[66] Practically, this involvement of mercenaries in the political sphere of war, by involving them in the decision making process as security experts, means that they are 'directly involved in establishing the priorities' of war and warfare, and mercenaries thus move beyond being engaged solely in the conduct of war to being 'part of the process of defining security

63 Anna Leander, 'The Power to Construct International Security: On the Significance of Private Military Companies', *Millennium: Journal of International Studies* Vol. 33, no. 3 (June 2005): 814.
64 Anna Leander, 'The Paradoxical Impunity of Private Military Companies: Authority and the Limits to Legal Accountability', *Security Dialogue* Vol. 41, no. 5 (October 2010): 474.
65 Anna Leander, 'Regulating the Role of Private Military Companies in Shaping Security and Politics', in *From Mercenaries to Market: The Rise and Regulation of Private Military Companies*, by Chesterman, Simon and Lehnardt, Chia (Oxford: Oxford University Press, 2007), 49; Leander, 'The Paradoxical Impunity of Private Military Companies', 472–74; Åse Gilje Østensen, 'In the Business of Peace: The Political Influence of Private Military and Security Companies on UN Peacekeeping', *International Peacekeeping* Vol. 20, no. 1 (February 2013): 34; Harold D. Lasswell, 'The Garrison State', *American Journal of Sociology* Vol. 46, no. 4 (1941): 455–68.
66 Deborah Avant, 'The Emerging Market for Private Military Services and the Problems of Regulation', in *From Mercenaries to Market: The Rise and Regulation of Private Military Companies*, ed. Simon Chesterman and Chia Lehnardt (Oxford: Oxford University Press, 2007), 189.

concerns and political priorities'.[67] Consequently, by leveraging their authority, which polities reaffirm when hiring them, mercenaries can affect 'strategies and priorities' that not only shape wider security discourses, but reach into the inner working of a polity's security priorities.[68] This fear of reliance on mercenaries for not only security functions, but also security understandings, stems from the assumption that knowledge of security *should* rest within the 'state'. However, as Parrott argues, with reference to the early modern mercenaries, a 'public-private partnership' was crucial in the development of the modern 'fiscal-military state'.[69] The dependence on mercenaries as 'specialists of violence' with a specific set of skills in war and warfare is therefore not a new phenomenon.[70]

Secondly, it has been argued that with the modern use of mercenaries, the 'market logic' that these actors work under, due to their 'private' nature, affects the demand and supply of security functions in a negative way. As Leander and van Munster explain, 'the market for 'security' generates its own demand insofar as the social perception of threat is both cause and effect of the marketing of security solutions'.[71] The logic is that, as private actors, mercenaries are fundamentally loyal and owe allegiance to the stakeholders that have an economic interest in their financial success, such as investors, rather than the hiring polity. This ultimately means, according to Leander, that 'threats and security concerns that might not previously have existed', are either exaggerated or made up, to maximise the hiring and earnings potential for mercenaries, skewing the knowledge of security and military functions in favour of private actors.[72] By manipulating 'public threat assessment and policy', mercenaries can create a 'demand for their service'.[73] Likewise, Krahmann highlights how mercenaries have created their own demand by continuously identifying and providing services to ever-evolving risks and in the process creating

67 Leander, 'Regulating the Role of Private Military Companies in Shaping Security and Politics', 52.
68 Østensen, 'In the Business of Peace', 35.
69 David Parrott, *The Business of War: Military Enterprise and Military Revolution in Early Modern Europe.* (Cambridge: Cambridge University Press, 2012), 317.
70 Michael Mallet, ed., 'Mercenaries', in *Medieval Warfare: A History*, by Maurice Keen (Oxford: Oxford University Press, 1999), 214.
71 Anna Leander and Rens van Munster, 'Private Security Contractors in the Debate about Darfur: Reflecting and Reinforcing Neo-Liberal Governmentality', *International Relations* Vol. 21, no. 2 (June 2007): 204–5.
72 Leander, 'The Power to Construct International Security', 814.
73 Elke Krahmann, 'Security: Collective Good or Commodity?', *European Journal of International Relations* Vol. 14, no. 3 (1 September 2008): 391.

a dependence on their services to combat the 'ontological insecurity' felt by their clients.[74]

The consequentialist moral objections in this sub-category are therefore focused on the negative consequences employing mercenaries will have on the *knowledge* surrounding security functions, but also on a fear that these actors will create their own demand. This was a fear that was shared by Machiavelli, when he stipulated that mercenaries 'cannot be trusted because they will always be trying to increase their own authority'.[75] In this speech act, with explicit reference to these actors' epistemic power, Machiavelli de-legitimised mercenaries based on the negative consequences their usage will have on the polity. As aforementioned, there were conflicting conceptions of the ideal polity in the Italian city-states, and the mercenaries were seen by many as illegitimate fighters disrupting the ideal polity.[76] Consequently, the moral objections in this sub-category are made towards mercenaries based on a fear that these actors' epistemic power will outweigh that of the polity. The foundation of this sub-category is therefore that the hiring polity reaffirms mercenaries' epistemic power simply by using them, and that this practice happens with detrimental consequences to the polity's ability to identify and respond to security issues, a function and responsibility that should rest with the polity.

Conclusion

The extensive use of mercenaries throughout history has been followed by questions and concerns about these fighters' legitimacy. Operating in a grey-zone of warfare, these actors have often been a vital part of war, employed by myriads of different polities, and their contribution to various kinds of warfare has been remarkable. However, as the historical examples show, alongside mercenaries' extensive usage has been the continuous construction of them as illegitimate fighters. This negative perception of mercenaries and the tendency to regard them with suspicion has a long history; or, as Sarah Percy has argued, 'for as long as there have been mercenaries, there has been a norm against mercenary use'.[77] Understanding

74 Elke Krahmann, 'The Market for Ontological Security', *European Security* Vol. 27, no. 3 (3 July 2018): 364.
75 Machiavelli, *The Prince*, 34.
76 Bayley, *War and Society*, 178.
77 Percy, *Mercenaries*, 1.

the historical context surrounding the de-legitimation of mercenaries can foster a deeper understanding of contemporary attitudes towards mercenaries, and by appreciating that resistance towards mercenaries and the active de-legitimation of them as fighters has been a common practice, especially in the advent of challenges to conceptions of the ideal polity, the future utility of these actors can be discussed on a more informed and nuanced basis.

Ultimately, the de-legitimation of mercenaries has happened via distinct discursive practices – moral objections. These moral objections can be divided into two main categories – deontological and consequentialist. These focus on the inner working of the mercenary and the consequences of using mercenaries for the hiring polity, respectively. The deontological moral objections made towards mercenaries have as their foundation the idea that mercenaries are illegitimate fighters due to their foreignness and their perceived monetary motive. As actors driven by selfish monetary gain, mercenaries are unable to dedicate themselves to a proper cause, and, without being driven by a patriotic sense of duty, mercenaries will be unable to make the ultimately communal sacrifice on the battlefield. The consequentialist moral objections made towards mercenaries, on the other hand, focus on the possible consequences that hiring mercenaries will have for the hiring polity. Within this category lies fears over mercenaries' inherent outsider status and how this status influences, infringes, and erodes core aspects of the polity. The moral objections made towards mercenaries ultimately rest upon a normative judgement of who *should* fight in war, and they rest on the assumption that there is a right and a wrong, a legitimate and illegitimate, way of being and doing when participating in war. The moral objections articulate this assumption and they therefore act as the foundation for the perception that mercenaries are illegitimate fighters. Mercenaries are not inherently illegitimate, they are made illegitimate via distinct discursive practices, and, in this way, the moral objections made against mercenaries both constitute and perpetrate the perception that mercenaries are illegitimate fighters. The moral objections made towards mercenaries have been strongest in situations where these actors have been perceived to challenge the predominate conceptions of the ideal polity. In situations where they have been viewed as disruptors of the ideal polity because they, as illegitimate fighters, offer the opportunity to re-evaluate how the ideal polity should be constructed and organised.

References

Alexandra, A. 'Private Military and Security Companies and the Liberal Conception of Violence'. *Criminal Justice Ethics* 31, Vol. no. 3 (2012): 158–74.

Austin, J. L. *How to Do Things with Words*. Cambridge, MA: Harvard University Press, 1975.

Avant, D. 'The Emerging Market for Private Military Services and the Problems of Regulation'. In *From Mercenaries to Market: The Rise and Regulation of Private Military Companies*, edited by Simon Chesterman and Chia Lehnardt. Oxford: Oxford University Press, 2007.

Avant, D. and L. Sigelman. 'Private Security and Democracy: Lessons from the US in Iraq'. *Security Studies* Vol. 19, no. 2 (2010): 230–65.

Barnett, C. *Britain and Her Army: A Military, Political and Social History of the British Army, 1509 – 1970*. London: Cassell & Co., 2000.

Bayley, C. C. *War and Society in Renaissance Florence: The de Militia of Leonardo Bruni*. Toronto: University of Toronto Press, 1961.

Bourdieu, P. *Language and Symbolic Power*. Cambridge: Polity Press, 1991.

Brauer, J. and H. van Tuyll. *Castles, Battles, and Bombs: How Economics Explains Military History*. Chicago & London: The University of Chicago Press, 2008.

British Foreign and Commonwealth Office. 'Private Military Companies: Options for Regulations'. Green Paper. London: House of Commons, 2002.

Burmester, H. C. 'The Recruitment and Use of Mercenaries in Armed Conflicts'. *The American Journal of International Law* Vol. 72, no. 1 (1978): 37–56.

Carmola, K. *Private Security Contractors and New Wars: Risk, Law and Ethics*. London: Routledge, 2010.

Chen, D. H. 'Holding 'Hired Guns' Accountable: The Legal Status of Private Security Contractors in Iraq'. *Boston College International & Comparative Law Review* Vol. 32 (2009): 101–13.

Coady, C. A. J. 'Mercenary Morality'. In *International Law and Armed Conflict*, by A.G.D. Bradney, 55–70. Stuttgart: Franz Steiner Verlag, 1992.

Congressional Records. House Adopts MEJA Expansion and Enforcement Act of 2007 (2007). https://fas.org/irp/congress/2007_cr/h-meja.html.

———. Military Extraterritorial Jurisdiction Act of 2000, Pub. L. No. 106–523 (2000). https://www.congress.gov/bill/106th-congress/senate-bill/768/text.

Fowler, K. *Medieval Mercenaries*. Vol 1: The Great Companies. Oxford: Blackwell Publishers, 2001.

France, J. 'Introduction'. In *Mercenaries and Paid Men: The Mercenary Identity in the Middle Ages*, by John France. Leiden: Brill, 2008.

Human Rights First. 'State of Affairs: Three Years After Nisoor Square'. Human Rights First, September 2010.

International Committee of the Red Cross. 'Protocol Additional to the Geneva Conventions of 12 August 1949, and Relating to the Protection of Victims of International Armed Conflicts (Protocol I)'. Article 47, 8 June 1977.

International Stability Operations Association. ISOA Code of Conduct, § Version 13.1. Accessed 31 May 2019. https://stability-operations.org/page/Code.

Joenniemi, P 'Two Models of Mercenarism: Historical and Contemporary'. *Instant Research on Peace and Violence* Vol. 7, no. 3/4 (1977): 184–96.

Kaldor, M. *New and Old Wars: Organized Violence in a Global Era*. Third Edition. Cambridge: Polity Press, 2012.

Keen, M. H. *The Laws of War in the Late Middle Ages*. London: Routledge & Keegan Paul, 1965.

Krahmann, E. 'From 'Mercenaries' to 'Private Security Contractors': The (Re Construction of Armed Security Providers in International Legal Discourses'. *Millennium: Journal of International Studies* Vol. 40, no. 2 (2012): 343–363.

———. 'Security: Collective Good or Commodity?' *European Journal of International Relations* 14, no. 3 (1 September 2008): 379–404.

———. 'The Market for Ontological Security'. *European Security* 27, no. 3 (3 July 2018): 356–73.

Krieg, A. 'Towards a Normative Explanation: Understanding Western State Reliance on Contractors Using Social Contract Theory'. *Global Change, Peace & Security* Vol. 25, no. 3 (1 October 2013): 339–55.

Lasswell, H. D. 'The Garrison State'. *American Journal of Sociology* Vol. 46, no. 4 (1941): 455–68.

Leander, A. 'Globalisation and the Eroding State Monopoly of Legitimate Violence'. Working Papers (Copenhagen Peace Research Institute (COPRI)). Copenhagen, Denmark: Copenhagen Peace Research Institute, 2001.

———. 'Regulating the Role of Private Military Companies in Shaping Security and Politics'. In *From Mercenaries to Market: The Rise and Regulation of Private Military Companies*, by Chesterman, Simon and Lehnardt, Chia. Oxford: Oxford University Press, 2007.

———. 'The Paradoxical Impunity of Private Military Companies: Authority and the Limits to Legal Accountability'. *Security Dialogue* Vol. 41, no. 5 (October 2010): 467–90.

———. 'The Power to Construct International Security: On the Significance of Private Military Companies'. *Millennium: Journal of International Studies* Vol. 33, no. 3 (June 2005): 803–25.

Leander, A. and R. van Munster. 'Private Security Contractors in the Debate about Darfur: Reflecting and Reinforcing Neo-Liberal Governmentality'. *International Relations* 21, no. 2 (June 2007): 201–16.

Liu, H-Y. and C. Kinsey. 'Challenging the Strength of the Antimercenary Norm'. *Journal of Global Security Studies* Vol. 3, no. 1 (1 January 2018): 93–110.

Machiavelli, N. *The Prince*. Edited by Robert M. Adams. 2nd ed. New York & London: Norton & Company, 1992.

Mallet, M. ed. 'Mercenaries'. In *Medieval Warfare: A History*, 209–30. Oxford: Oxford University Press, 1999.

———. *Mercenaries and Their Masters: Warfare in Renaissance Italy*. London: Bodley Head, 1974.

Mandrup, T. 'Privatisation of Security: The Concept, Its History and Its Contemporary Application'. Copenhagen, Denmark: Royal Danish Defence College, 2012.

McGrade, P. 'Private Military and Security Companies'. *Chatham House: International Law Summary*, 2013, 10.

Minow, M. 'Outsourcing Power: How Privatizing Military Efforts Challenges Accountability, Professionalism, and Democracy'. *Boston College Law Review* Vol. 46, no. 5 (5) (2005): 989–1026.

Mockler, A. *Mercenaries*. London: Macdonald, 1969.

Morton, J. S. and P. Jones. 'The Legal Status of Mercenaries'. *Politics & Policy* Vol. 30, no. 4 (2002): 624–45.

O'Brien, K. 'What Should and What Should Not Be Regulated?' In *From Mercenaries to Market: The Rise and Regulation of Private Military Companies*, by Simon Chesterman and Chia Lehnardt. Oxford: Oxford University Press, 2007.

Østensen, Å. G. 'In the Business of Peace: The Political Influence of Private Military and Security Companies on UN Peacekeeping'. *International Peacekeeping* Vol. 20, no. 1 (February 2013): 33–47.

Parrott, D. *The Business of War: Military Enterprise and Military Revolution in Early Modern Europe*. Cambridge: Cambridge University Press, 2012.

Pattison, J. *The Morality of Private War: The Challenges of Private Military and Security Companies*. Oxford: Oxford University Press, 2014.

Percy, S. *Mercenaries: The History of a Norm in International Relations*. Oxford: Oxford University Press, 2007.

———. 'Regulating the Private Security Industry'. Adelphi Paper Vol. 384. London: Routledge & IISS, 2006.

Phelps, M. L. 'Doppelgangers of the State: Private Security and Transferable Legitimacy'. *Politics & Policy* Vol. 42, no. 6 (1 December 2014): 824–49.

Potter, K. R. trans. *Gesta Stephani*. London: Nelson, 1955.

Riemann, M. "As Old as War Itself'? Historicizing the Universal Mercenary'. *Journal of Global Security Studies*, 26 February 2020, 1–16.

Searle, J. R. *Speech Acts: An Essay in the Philosophy of Language*. Cambridge: Cambridge University Press, 1969.

Singer, P.W. *Corporate Warriors: The Rise of the Privatized Military Industry*. Ithaca, N.Y.: Cornell University Press, 2008.

Tavernise, S. 'U.S. Contractor Banned by Iraq Over Shootings'. *The New York Times*, 18 September 2007. https://www.nytimes.com/2007/09/18/world/middleeast/18iraq.html.

The Parliamentary History of England, from the Earliest Period to the Year 1803. Vol. 18 (1774-1777). London: Printed by T.C. Hansard, 1813.

Thomson, J. E. *Mercenaries, Pirates, and Sovereigns: State-Building and Extraterritorial Violence in Early Modern Europe*. Princeton, N.J.: Princeton University Press, 1994.

United Nations Commission on Human Rights. 'Report of the Working Group on the Use of Mercenaries as a Means of Violating Human Rights and Impeding the Exercise of the Right of Peoples to Self-Determination'. A/HRC/36/47, 20 July 2017.

Weber, M. *From Max Weber: Essays in Sociology*. London & New York: Routledge, 2013.

3

THINKING BEYOND THE DIVIDE

The Complex Relationships Between State/Non-State Armed Actors in Syria's War

Helle Malmvig

Introduction

As the tide of war has turned in Bashar al Assad's favour, the Syrian President claimed frequently that the government was successfully reasserting control over the areas previously governed by Syrian rebel groups.[1] Thus, when US Forces partially pulled out of North Eastern Syria in October 2019 and Assad and Russian forces rolled in, the Syrian President could once again claim that Damascus was about to establish full sovereignty over every inch of Syrian territory, including the Kurdish dominated Syrian Democratic Forces (SDF).[2]

1 Syrian President Bashar al-Assad asserted that the Syrian army's deployment was not only for security or military purposes: 'The deployment of the Syrian army is an expression of the presence of the Syrian state, which means the presence of all the services that should be provided by the state'. Interview given to Al-Souriya and Al-Ikhbariya TV, November 1, 2019. Quoted from International Crisis Group (ICG). 'Steadying the New Status Quo in Syria's North East'. *Briefing* 72 (Brussels: ICG, 2019), 4.

2 Damascus has made clear that it intends to reassert full state control over the north east, saying individual SDF members could join the Syrian army. Interview to SANA TV, 31 October quoted from ICG, 'Steadying the New Status Quo', 4.

The Syrian war has from early on been a battle of competing narratives. This is also true when it comes to the nature and supposed 'strength' of the Syrian state. Indeed, many in the Syrian opposition, as well as scholars of Syrian politics, argue that although the Assad government may have won victories on the battlefield, the regime in practice exercises little control over its territory and the plethora of armed militias supposedly loyal to it. The many competing armed factions, foreign militias and mercenaries are in fact hollowing out the state's power from within, and Bashar al Assad is merely presiding over a hollow shell. Syria has, it is argued, become a 'failed state' and Assad a mere 'King of Ashes'.[3]

The above two narratives on the nature of Syrian statehood may seem contradictory. But they in fact build on similar and prevalent assumptions about what a state is – exclusive territorial control and monopoly on means of the violence – and they both presume that Violent Non-State Actors (VNSAs) necessarily undermine and challenge that very 'stateness', that civil war, in the words of Staniland essentially is a 'struggle over the monopoly of violence'.[4] This chapter will problematize such presumptions, showing how the Syrian state in fact engages in very complex overlapping authority relationship with VNSAs, at times even ruling through these, at others being enmeshed in intersecting nodes of authorities, where multiple and shifting armed groups overlie and substitute the Syrian government, and that these are not only over means of violence but also over authority and governance. The Syrian regime has in practice outsourced security to a range of foreign and some loyalist militias in order to survive, often exercising limited command and control over these. Similarly, while rebel forces have battled the Syrian government forces (SAA), they have also carried out a range of state tasks in lieu of the Syrian government, frequently operating within existing Syrian state laws. And finally, while the Kurdish Yekîneyên Parastina Gel (YPG) for instance has taken care of policing and security in the North East, with little intervention from the Syrian state, they have also loosely coordinated governance with the regime and allowed it a scattered physical presence. The boundary between state and non-state is therefore far from a zero-sum struggle over the monopoly of violence.

3 Christopher Phillips, The Battle for Syria: International Rivalry in the New Middle East. (New Haven: Yale University Press, 2016), 2; Raymond Hinnebusch, After the Arab uprisings: between democratization, counter-revolution and state failure. (London: Routledge, 2016), 560; Sarah Hunaidi. 'ISIS has not been defeated. It's alive and well in Southern Syria'. *Foreign Policy*, April 3, 2019.

4 e.g. Anthony Vinci. 'Anarchy, failed States and Armed Groups: Reconsidering Conventional Analysis'. *International Studies Quarterly*, 52:2 (2018).

Armed non-state actors do not necessarily undermine or take something away from the state but may complement and add to the state; the state in practice delegating certain types of violence, or governance to VNSA's, or the state tactically cooperating with specific VNSA's in specific geographical or functional areas. The boundary between VNSA's and the state is often so fluid that what constitute state and non-state forces is continuously contested and performed by the many armed actors themselves.[5]

To analyze these complex and grey-zone relationships between state and non-state armed actors in Syria, the chapter proposes a typology of three forms of interactions drawing partially on Staniland's seminal typology of state-insurgency relations and insights from recent literature on rebel governance and hybrid warfare.[6] Relations between the Syrian state and VNSA's are thus analytically structured in to three types, these being i) alignment, ii) indifference/tactic cooperation, and iii) hostility.[7] These three types also serve to structure the analysis in three main parts: part one will focus on VNSAs that are closely aligned or partially embedded with Syrian government forces, the second part on state-non-state relations marked by indifference or what Staniland calls passive cooperation, and the third part on VNSA's that primarily have hostile relations with the state. Each section will also analyze how relations have developed over time from the early years of the uprising in 2011 until 2019, in sum showing how the Syrian state is far from a centralized Leviathan that is gradually regaining control, but instead deeply entangled with and dependent on armed non-state actors and reversely how these VNSA do not merely hallow out the state, but may add and perform the state, or simply entertain rather passive tactic relations.

5 See also Paul Staniland. 'States, Insurgents, and Wartime Political Orders'. *Perspectives on Politics*, 10:2 (2012); Paul Staniland. 'Militias, Ideology, and the State'. *Journal of Conflict Resolution*, 59:5 (2015); Vladimir Rauta. 'Towards a typology of non-state actors in hybrid warfare: proxy, auxiliary, surrogate and affiliated forces'. *Cambridge Review of International Affairs*, September 9, 2019, available at https://doi.org/10.1080/09557571.2019.1656600.

6 Staniland, 'Militias, Ideology, and the State'.

7 In addition to secondary literature, the chapter is based on primary research in Syria's neighboring countries (Turkey, Lebanon, Jordan, Iraq, and Iran). More than 60 diplomats, activists, intellectuals, militia leaders (foreign and Syrian), and Syrian fighters have been interviewed in person or contacted remotely, over the last 8 years of the war. For reasons of security they have all been anonymized.

Mosaics of Power: State and Non-State Relations in Syria[8]

Over the last decade several scholars working on rebel governance, insurgency and hybrid security/hybrid war in the Global South have shown how non-state armed groups relate in manifold and complex ways to the state. Armed non-state actors, as emphasized above, may not only act in opposition to the state, but may act on behalf of the state, in cooperation with a state's regular armed forces, or with a state's privately contracted paramilitaries. Even terrorist groups may have informal ad hoc relations with the state or intermediaries of the state, just as armed criminal networks often clandestinely do.[9] Non-state actors are in short, despite their name, often deeply involved with state power, and non-state-state relations can take multiple complex forms that range from rivalry, to indifference, outsourcing and the delegation of power.

Staniland's three-fold typology of state-insurgent relations/corporation (active, passive, non-existent) is close to the one adopted here, yet it does not account for armed non-state actors who are neither insurgents nor part of a state's regular forces, such as the omnipresent foreign Shia militias fighting alongside Syrian government or the many civil defence forces only loosely connected to the state. Syria to some extent resembles other hybrid battlefield such as Ukraine, where state parallel militias work as what Vladimir Rauta in a recent typology respectively calls auxiliary, proxy, surrogate and affiliated forces.[10] Yet in order to keep the typology simple, and still add a 'non-insurgent category', the typology used here consist of three main forms of relations between state and armed non-state actors in the Syrian context: 1) state-aligned groups, where relations are primarily guided by delegation, cooperation or toleration; 2) groups, where relations with the state are guided by tactical indifference or ad-hoc

8 Throughout this chapter, the terms Assad government, Assad regime and Syrian state (understood as state institutions in a Weberian sense) are used interchangeably, in so far as Syrian state institutions cannot meaningfully be said to operate independently from the Assad regime, who 'captured' these very institutions after the coup d'état in 1970.

9 See e.g. Keith Krause and Jennifer Milliken. 'The Challenge of Non-State Armed groups'. *Contemporary Security Policy*, 30:2 (2009), 202-220; Eran Zohar. 'A New Typology of Contemporary Armed Non-State Actors: Interpreting the Diversity Studies in Conflict and Terrorism'. *Studies in Conflict and Terrorism*, 39:5 (2016), 423-450; Diane Davis. 'Non-State Armed Actors, New Imagined Communities and Shifting patterns of Sovereignty and Insecurity in the Modern World'. *Contemporary Security Policy*, 30:2 (2009), 221-245; Jutta Bakonyi and Kirsti Stuvøy. 'Violence and Social Order beyond the State: Somalia and Angola'. *Review of African Political Economy*, 32:104 (2005), 359–382; Klaus Schlichte. *In the Shadow of Violence: The Politics of Armed Groups*. (Frankfurt & New York, NY: Campus Verlag, 2009).

10 Rauta, 'Towards a typology of non-state actors in hybrid warfare'.

cooperation; 3) rebel groups, where relations are guided by competition and hostility. Inevitably, these three types are heuristic devices, as state-non-state relations are fluid and shifting over time and locales. Some groups may fall between categories, just as the same group may move from one to another. The Islamic State for instance tactically cooperated with the Syrian regime in the years 2013-2015 falling into category two, while in 2019 it clearly falls into category three. Similarly, transnational Shia militias are formally aligned with the Assad government, i.e. category one, but have occasionally clashed with regular Syrian army forces around the Iraqi-Syrian border crossing in the eastern part of Syria, thus partially falling into category three. In this sense, state-non-state relations and alliances shifted back and forth over time, showing the importance of changing interests and strategies in civil wars rather than just raw capacity.[11]

State-Aligned Armed Groups

Early in the conflict the Assad regime came to rely on a range of irregular forces only loosely connected to the state in order to guarantee its survival. First, it was mainly citizen militias and criminal networks, and later increasingly also foreign groups such as the Lebanese Hezbollah, Shia fighters recruited by Iran from Pakistan, Afghanistan and elsewhere, and even privately contracted soldiers from the Russian Wagner Group.[12] Common for this wide array of armed non-state forces, as will be shown below, are that they were aligned with the Syrian state, their use of violence being either informally state sanctioned or, at a minimum, implicitly tolerated by the state. This, however, does not imply that they were subordinated to, or controlled by the Assad government. Rather, these armed groups often acted rather autonomously, and at times even competed and fought with the regular forces or each other.

As the uprising gained momentum in early 2011 and mass demonstrations broke out all across Syria, the Assad government, fearing its very survival was at stake, soon engaged its most loyal security forces

11 Staniland, 'States, Insurgents, and Wartime Political Orders', 249.

12 See also Reinoud Leenders and Antonio Giustozzi. 'Outsourcing state violence: The National Defence Force, 'stateness' and regime resilience in the Syrian war'. *Mediterranean Politics*, 24:2 (2017); al-Tamimi, 'The Story of al-Hajj Azra'il: From Nubl to Deir az-Zor', *Pundicity*, March 16, 2018, available at http://www.aymennjawad.org/2018/03/the-story-of-al-hajj-azrail-from-nubl-to-deir-az.

and called upon armed civilian gangs – *shabihas*[13] – to clamp down on demonstrators and dissent.[14] *Shabiha* networks with close ties to the regime ruthlessly pursued demonstrators and known activist from the emerging opposition by carrying out assassinations and intimidations. Later, *shabihas* in black irregular uniforms supported the regular army's fourth division in assaults in several Syrian villages. *Shabihas* were also involved in targeted killings including a sectarian massacre in Houla.[15] The regime employed the *shabihas* to act in the shadows, dividing the population by stoking up sectarian fears, especially with minority groups.[16] In the early stages of the uprising, the *shabiha* thus enabled the regime to act with impunity and brutality by way of a non-state proxy. The Assad government could publicly claim that it was not involved in atrocities and sectarian divide-and-rule tactics,[17] and could use forces that it knew would be loyal to the regime.

Over the course of 2012, the uprising evolved into mass state repression and outright war. The Free Syrian Army (FSA) was gradually formed by defecting government soldiers; later, civilian opposition members also joined.[18] Much needed arms and financing came from governments and business leaders in the Gulf and Turkey. Similarly, the Assad government's main allies Iran and Hezbollah, fearing that the Assad regime was about to fall, and that weapons transfers to Hezbollah in Lebanon against Israel would be disrupted, sent military advisors and elite forces to Syria. Iran in particular came to dominate the training, financing and equipping of Syrian

---

13 The term *shabiha* originates from the Arab word *Shabh* meaning ghosts, they first appeared in the 1970s under President Hafez al Assad where mafia and criminal networks with close ties to the Assad regime and the paramilitary Defence Brigades, were cracking down on opposition and were carrying out protection racketeering, especially in the coastal areas.

14 Christopher Phillips. 'Syria after IS'. *Orient,* 4 (2017), 16-23. Steven Heydemann. 'Beyond Fragility: Syria and the Challenges of reconstruction in Fierce States'. *Report* (Washington DC: Brookings Institute, 2018).

15 BBC. 'Syria unrest: Who are the Shabiha?' *BBC News,* May 29, 2012; ICG. 'Syria's Mutating Conflict'. *Briefing 128* (Brussels: ICG, 2012).

16 Christopher Phillips. 'Sectarianism as Plan B' in *Saudi Arabia and Iran: The Struggle to Shape the Middle East,* ed. Simon Mabon (London: Foreign Policy Centre, 2018).

17 See e.g. ICG. 'Popular Protest in the Middle East and North Africa: the Syrian Regime's Slow-motion Suicide'. *Report 109,* July 13, 2011 (Brussels: ICG, 2011); Geraint Alun Hughes. 'Syria and the Perils of Proxy Warfare'. *Small Wars and Insurgencies,* 25:3 (2014); Helle Malmvig. 'Coming in from the Cold: How we make take sectarianism seriously without playing to the tunes of regional power elites'. *International Relations Theory and a Changing Middle East Study 16,* September 17, 2016 (Washington DC: Project on Middle East Political Science, 2015).

18 Samer Abboud. 'Hard Road ahead for the Syrian Exile Government'. November 15, 2013 (Beirut: Carnegie Middle East Center, 2013); Helle Malmvig. 'Eyes Wide Shut: Power and Creative Visual *Counter*-Conducts in *the Battle for Syria'. Global Politics,* 30:2 (2016); Phillips, *The Battle for Syria.*

local militias and/or sectarian transnational groups, while Hezbollah mainly contributed with its own fighting forces and with training of local Syrian militias.

Inside Syria the *shabihas* morphed into hundreds of locally based citizen militias (known as Popular Committees). These were only partially armed by the government and often operated rather autonomously, partaking in criminal activities and local power games.[19] By 2013 Iran, increasingly impatient with their unruly behavior, spearheaded attempts to integrate all of the various militias into a more effective para-military force structure modelled on the Popular Mobilization Units's in Iraq or the Hezbollah in Lebanon.[20] These came to operate under broad umbrella terms known as the Local Defence Forces (LDF) and, later, National Defence Forces (NDF) by making use of a paragraph in Syrian military law that allowed for auxiliary forces to supplement the regular army.[21]

Foreign influence held sway over most of these irregular forces from the beginning, as these were predominantly trained and financed by the Iranian Revolutionary Guard (IRGC) and Lebanese Hezbollah, and later Russia.[22] Some of the militias, such as the Fatimiyun, al Nujaba and Liwa Abu Fadl al Abbas, were entirely made up of foreign Shia fighters from Iraq, Pakistan and Afghanistan. Other militia groups such as Liwa al Baquir were created and run by Hezbollah and Iran, although the rank and file comprised local Syrians. There were also militia groups more closely linked to the regime's inner circle in terms of their finance and control, such as Liwa Imam al Baqr and the Tiger Force, both run by the Syrian intelligence agencies, and Fahud Homs, Suqur Sahra Dir and the Fourth and Fifth Volunteer Assault Corps run respectively by the infamous regime oligarchs Rami Makhlouf, Ayman Jaber and Maher al Assad.[23] In addition, there are LDF militias linked to political parties such as the Baath Battalions and the Syrian Socialist National Party or to the tribes in Deir Ezzour and Suwayda.[24]

---

19 Aron Lund. 'Syrian Jihadism'. *Swedish Institute of International Affairs*, September 14, 2012.
20 Tim Eaton et al. 'Conflict Economies in the Middle East and North Africa'. *Report*, June 25, 2019 (London: Chatham House, 2019), 15.
21 Lina Khatib and Lina Sinjab. 'Syria's Transactional State: How the Conflict Changed the Syrian State's Exercise of power'. *Research Paper*, October 10, 2018 (London: Chatham House, 2018) 15.
22 Haid Haid. 'Reintegrating Syrian Militias: Mechanisms, Actors and Shortfalls'. December 12, 2018 (Beirut: Carnegie Middle East Center, 2018).
23 Abdulrahman al Masri. 'Analysis: The Fifth Corps and the State of the Syrian Army'. *Syria Deeply*, January 11, 2017.
24 See Helle Malmvig. 'Mosaics of Power: Fragmentation of the Syrian State since 2011'. *Report* (Copenhagen: Danish Institute for International Studies, 2018), 4.

These irregular militia groups are not only fighting and providing much needed manpower, but they are manning checkpoints, policing the streets and engaged in extraction and protection racketeering at times with little or no central command from the Syrian government or the regular forces.[25] Pro-regime militias for instance set up check points solely for extracting money, rent out check points to private individuals, or offer fake security identity cards for those who wish to avoid military service and are able to pay. As Khatib and Sinjab note 'the regime tolerates this activity not only because it does not have the financial means to pay all of its affiliates, but also because allowing such groups to profiteer from the war is a way of keeping them on its side'.[26] Given years of accelerated war-economy and international sanctions, there is a close interconnectedness between the means of violence and the economy, which at times have led to open clashes and tension between regular forces and the militias formally aligned with the government. Pro-regime militias have, for example, fought for control over checkpoints, lucrative smuggling routes or the terms of local ceasefires.[27]

Yet the informal delegation of force and governance to local militia-business networks has also allowed the government to keep and develop patronage networks that remain relatively dependent on the Assad government. Militia leaders, affiliated businessmen and close relatives act as the heads of charities and so-called Non-Governmental Organisations (NGOs) that take care of martyrs' families, rebuild schools and even run some public transport.[28] Local business leaders who help fund such armed groups are given rewards and remunerations in return for these services, such as government positions in the public sector or with local intelligence bureaus. They may also be appointed as sheiks or heads of one of the many new charitable organizations that act as intermediaries for the regime. All of this allows the government to maintain some degree of surveillance over and penetration of civil society. In terms of the latter for instance, to

---

25 See e.g. Leenders and Giustozzi, 'Outsourcing state violence'; Aron Lund. 'Blame Game over Syrians Stranded in the Desert'. *Report*, June 18, 2018 (Washington DC: The Century Foundation, 2018); al-Tamimi, 'The Story of al-Hajj Azra'il'.

26 Khatib and Sinjab, 'Syria's Transactional State', 19.

27 ICG. 'Avoiding a Free-for-all in Syria's North east'. *Briefing 66*, December 21, 2018 (Brussels: ICG, 2018); Haid Haid. 'HTS's Offline Propaganda: Infrastructure, Engagement and Monopoly'. *Report*, September 18, 2019 (London: International Centre for the Study of Radicalisation, 2019); Eaton et al. 'Conflict Economies'.

28 Kheder Khaddour. 'Strength in Weakness: The Syrian Army's Accidental Resilience'. March 16, 2016 (Beirut: Carnegie Middle East Center, 2016).

access international funding and aid, notably from the United Nations, larger NGOs need to register with the Syrian Foreign Ministry, which often requires connections and bribes. To stay on the 'approved' list, charities are reportedly told to appoint regime-affiliated militia leaders and risk being removed from the list should they fail to comply.[29] Local businessmen may also serve as intermediaries in order to circumvent international sanctions and access funding from international donors. Similarly, business leaders have engaged in smuggling and trading in oil from, for example, Islamic State-controlled areas to the regime or in illegal trading from regime-held territory to besieged areas.[30] Big charities which fund the pro-regime militias and even pay the salaries of civil servants, are set up by the government's inner circle of oligarchs such as Mohammed Hamsho, George Haswani or Rami Maklouf's Bustan Organization. A number of decrees and laws[31] have also enabled the Syrian government to expropriate land and generate new funds through property development to compensate its new and old networks of business-militiamen.[32] Several Syrian interviewees point out that the crony capitalist order of illegal trading, transfer of public assets and licenses to business leaders closely associated with the Assad government that flourished before the war remain relatively intact, merely having intensified under the war economy.[33]

As the Syrian government began to regain lost territory from opposition groups, in particular after Russia's intervention in 2015, new attempts by the regime and Russia were made to integrate all of the local NDF and the LDF militias into the regular Syrian army structures. But these initiatives were allegedly met with strong Iranian and Hezbollah opposition.[34] At present the LDF is formally integrated with the Syrian army forces 4th division, but the strong Iranian influence remains.[35] According to several reports, fighters in the Syrian Army's 4th division, despite wearing its uniform and

29 Khaddour, 'Strength in Weakness'.

30 Aurora Sottimano. 'Building authoritarian 'legitimacy': domestic compliance and international standing of Bashar al-Asad's Syria'. *Global Discourse,* 6:3 (2016).

31 'The New Urban Renewal Law in Syria', *Syrian Law Journal,* May 14, 2018, available at http://www.syria.law/index.php/new-urban-renewal-law-syria/.

32 See e.g. Erwin Van Veen, 'Creating a new Syria: Property, Dispossession, and Regime survival', *Syria Comment,* April 15, 2018; Maha Yahya. 'What Will It Take for Syrian Refugees to Return Home?' *Foreign Affairs,* May 28, 2018.

33 See Sottimano, 'Building authoritarian 'legitimacy''.

34 Leenders and Giustozzi, 'Outsourcing state violence'; Haid Haid, 'Reintegrating Syrian Militias'.

35 Aymenn Jawad al-Tamimi. 'Administrative Decisions on Local Defence Forces Personnel: Translation & Analysis'. May 3, 2017, available at http://www.aymennjawad.org/2017/05/administrative-decisions-on-local-defence-forces; Haid Haid, 'HTS's Offline Propaganda'.

insignia, continue to be paid, trained, equipped by Lebanese and Iranian commanders and they are under their purview.[36] Hezbollah and Iran clearly favour this hybrid situation of overlapping and fragmented security orders, as this allows them to maintain influence and strategic depth even after the war, and thus remain embedded within the Syrian political system as in Iraq and Lebanon.[37] One Iranian analyst called Syria a situation of 'organized chaos',[38] while a senior Hezbollah official told this author that the commitment is rather open-ended 'Hezbollah will remain in Syria, as long as we are needed'.[39]

In sum, multiple pro-regime militias play a vital role in upholding the Assad government, substituting for and acting on behalf of the state in towns and cities that have slipped out of government control. But the relationship is not hierarchical, or one of formal command and control. Rivalry, turf war and foreign strategic interest co-exist, marking out competing orders of authority. Within this fragmented order so-called loyalist pro-regime forces are not necessarily loyal, and the boundary between state and non-state forces is so fluid that foreign militias or foreign sponsored militias are embedded within the regular force structure of the 4th and 5th Corp. This of course makes it difficult for the Assad government itself to smoothly translate victories on the battlefield by the many pro-regime forces into a strengthening of central government control that the regime so often claims to have gained.

Indifference and Tactical ad hoc Cooperation

Other armed-non state actors, neither fighting alongside the Assad government nor being state sanctioned, are nevertheless carrying out vital state and security tasks without the Assad government's interference. This applies to the Kurdish armed groups (YPG/YPJ)[40] in the North east, to some of the tribes and clans in Syria's East, and to a lesser extent, to the Islamic State, when it still held control of Syrian territory.

To understand how this relationship of tactical indifference or even ad hoc cooperation developed, it is necessary to go back to 2012 when Assad forces were severely pressured by opposition groups and the regime decided

36 ICG. 'Lessons from the Syrian State's return to the South'. *Report 196*, February 25, 2019 (Brussels: ICG, 2019a) 14; Haid Haid, 'HTS's Offline Propaganda', 2.
37 Malmvig, 'Mosaics of Power'.
38 Conversation in Tehran, November 2015.
39 Interview with senior Hezbollah Foreign Policy official Beirut, May 2018.
40 I.e. Yekîneyên Parastina Gel and Yekîneyên Parastina Jin (The People's Protection Units).

to pull most of its military and government officials out of Syria's northern regions dominated by Kurdish groups. Kurdish identity and politics had for decades been repressed under Assad rule, and Kurds had in particular been prone to persecution by the intelligence services *Mukhabarat*.[41] Now the regime's tactical withdrawal meant that it de facto handed over the territory to Kurdish groups and smaller competing militias, while saving resources and manpower for fighting elsewhere. Filling the void, Kurds set up self-governing local assemblies, police and military forces, and these assumed a range of government tasks such as health services, schools and security, eventually battling the Islamic State during its assault of towns and villages in Syria's north in 2013-2014.

Kurdish forces continue to cooperate with the US-led coalition in the fight against the Islamic State and have received weapons and crucial air-support within the framework of the Syrian Democratic Forces. But the YPG / YPJ has been careful never to attack regime forces, just as they have not been targeted by regime forces in return. Relations with the Syrian state have generally been marked by cautious neutrality or even ad hoc cooperation. In the main city of Qamishli for instance the regime controls the airport and a military base, just as the Assad government has paid the salaries of some state workers and civil servants. Some services such as higher education and transport have also been partially planned and coordinated with the Assad government, just as the government at times has provided official state documents such as passports and birth certificates.[42] In early 2019 the regime even set up a local council for issuing official Syrian state documents (Kurdish 24). This creates, as one Kurdish interviewee remarked to the author, a shadowy presence of the Assad state, and has worked as a constant reminder that the Assad government intends to reclaim its full authority in due course.[43] Yet at the same time, Kurdish local authorities continue to conduct crucial government tasks such as managing the informal border crossing with Iraq at Semalka-Faysh Khabour, where Kurdish forces control the passage of goods and people,[44] conducting internal police and security through the Kurdish Asayish and local governance structures. The Syrian government has in this sense *informally* outsourced several key tasks and

---

41 Jordi Tejel, Syrian Kurds: History, Politics and Society. (London: Routledge, 2009).
42 Kheder Khaddour. 'Local Wars and the Chance for Decentralized Peace in Syria'. March 28, 2017 (Beirut: Carnegie Middle East Center, 2017).
43 Author interview with Kurdish academic, August 2018.
44 See also Leila Vignal. 'The changing borders and borderlands of Syria in a time of conflict'. *International Affairs*, 93:4 (2017).

services normally associated with statehood, including control of the international border and policing to the YPG/YPJ, while maintaining a very light administrative and military footprint in the area. Despite of the Turkish incursion in October 2019 and partial withdrawal of US forces from the northeast this pattern appears to continue.[45]

Relations between Syria's many armed clan and tribal families and the Syrian state are similarly characterized by relative autonomy, fluidity and ad hoc cooperation. In the North East and Deir Ezzour alone, there are well over 70 armed tribes. These extended families are divided among themselves, as some are fighting alongside the Kurds within the SDF (e.g. the Shammar tribe), others are affiliated with the regime or Iran such as Tay, Akidat and the Shaitat tribes, while Turkey for its part has sponsored the formation of the 'Eastern Tribes Army' and Saudi Arabia 'the King Abdul Aziz brigade'.[46] Internally, generational divides often run deep. The younger generation has in general been more prone to join the Syrian opposition or the Islamic State, while elders have tended to hold on to state privileges and their local authority structures obtained by being historically loyal to the Assad family.[47] When the Islamic State ruled in Deir Ezzour and Raqqa some bedouin leaders publicized their loyalty to IS and were responsible for collecting taxes and continued carrying out civil documentation and arbitration.[48] When the tide of war turned, bedouin leaders reached out to the SDF or the regime. Similarly, tribes who, early in the war, were strongly connected to Saudi Arabia, such as the Akidat in Deir Ezzour, are now moving closer to the Assad government and Iran.[49] Over the course of the war, 'loyalties' have shifted and turned depending on who were the strongest powers on the ground.[50] External patrons may be paying salaries

45 In the wake of the US partial withdrawal the Syrian regime sent men and vehicles to the border post in the North especially around Kobani, but according to ICG so far has not aimed to take control of YPG held populated areas suggesting that the YPG administration may continue to govern in these areas. ICG, 'Steadying the new Status Quo'.

46 Haian Dukhan, *State and Tribes in Syria: Informal Alliances and Conflict Patterns*. (London: Routledge, 2018); Abdullah al-Ghadawi, 'New Configuration in the Syrian Regime's Security Structure'. *Analysis*, November 2019 (London: Chatham House, 2019).

47 See especially Haian Dukhan, 'Critical analysis of attempts to co-opt the tribes in Syria. Conflict Research Programme Blog', May 2, 2019, available at https://blogs.lse.ac.uk/crp/2019/05/02/critical-analysis-of-attempts-to-co-opt-the-tribes-in-syria/; Dukhan, *State and Tribes in Syria*; Tahir Zaman, *Islamic Traditions of Refuge in the Crises of Iraq and Syria*. (London: Palgrave Macmillan, 2016).

48 Dukhan, 'Critical analysis', 6.

49 Ghadawi, *New Configuration*.

50 Akil Hussein. 'Divisions defines Syria's tribes and Clans'. *Analysis*, January 2018 (London: Chatham House, 2018).

and arms, and tribal relations with the external powers, including the state, are highly fluid and marked by a strong sense of pragmatism and local autonomy. Tribal families maintain their own court systems, reconciliation mechanisms and local 'police'. Yet several Syrian interviewees with links to tribal groups have related that lawlessness reigns in many places.[51] Tribes are also heavily involved in criminal networks and smuggling routes across the border with Iraq; networks which the Islamic State and before that al Qaeda took use of, and tribal militias have been selling oil and gas to the regime as well as to the opposition.

Similarly, although relations between the Islamic State and the Assad government are hostile, both have also entertained ad hoc cooperative relations especially in the early phases of the war. The Syrian government has purchased oil and gas through intermediaries and even provided access to the national electricity grid to the terrorist group, thereby indirectly contributing to the financing and infrastructure of the Islamic State.[52] The Assad regime and its allies, Russia and Iran, infrequently targeted Islamic State positions. Instead they concentrated on attacking rebel groups associated with the Free Syrian Army.[53] Numbers vary, but there are estimates that up to 90 per cent of government-Russian air raids in 2015 targeted opposition positions and only 5 per cent Islamic State.[54] This allowed the Islamic State for a period to build relatively effective institutions and to mobilize fighters and local support structures vis-à-vis competing rebel groups. More than 15,000 pages of IS documents recovered from Mosul, show how the caliphate was running a network of police stations, prison, courts, ministries and higher education institutions, and how it sought to regulate even minuscule details of people's everyday lives from driving to fishing regulations.[55]

Tactically cooperating with jihadi terrorist groups was not entirely new to the Syrian government either. In the 2000s, the Assad government

51 Interviews with FSA fighters Jordan, September 2017 and Syrian journalist, Copenhagen, June 2019. See also 'A year into the Settlement: The Regime's fragile security and control rule the landscape'. *Enab Baladi*, June 30, 2019; Hunaidi, 'ISIS has not been defeated'.

52 See e.g. Hussein Almohamad and Andreas Dittmann. 'Oil in Syria between Terrorism and Dictatorship'. *Social Sciences*, 5:2 (2016), 1-16; Benoit Faucon and Ahmed Al Omran. 'Islamic State Steps Up Oil and Gas Sales to Assad Regime'. *The Wallstreet Journal*, January 19, 2017.

53 Michael Becker. 'When Terrorists and Target Governments Cooperate: The Case of Syria'. *Perspectives on Terrorism*, 9:1 (2015), 97.

54 Alina Polyakova, 'Russia Is a Great Power Once Again: Putin's Syria intervention saved Assad. But is he ready for what comes next?' *The Atlantic*, February 26, 2018; Jack Stubbs. 'Four-fifths of Russia's Syria strikes don't target Islamic State: Reuters analysis'. *Reuters*, October 21, 2015.

55 See e.g. Tom Stevenson 'How to run a Caliphate'. *London Review of Books*, June 20, 2019, 9.

had clandestinely sent Sunni militants to Iraq to fight the US forces, just as the Syrian regime released known terrorists from government prisons in the early days of the uprising, many of whom later became Islamic State and al Qaeda / al Nusra fighters.[56] This peculiar form of tactical relationship between an incumbent government and a terrorist group formally hostile to it is rare, but not unique.[57] In Algeria, for instance in the 1990s the military-ruled regime and its secret services both fought and collaborated with Islamist terrorists, turning a blind eye to the atrocities and massacres of the Armed Islamic Group of Algeria, even dressing up as terrorists in order to de-legitimize the more moderate Islamist opposition.[58]

Today in 2020, the Islamic State has been pushed out of its last pockets of territory, although some fighters and supporters remain. They are living underground in the Iraqi-Syrian border and desert areas, or have fled to Turkey, while others have been incorporated into the old tribal structures and Saddam Hussein's former intelligence networks.[59] The Islamic State may therefore be able to re-group and re-emerge in Syria, just as al Qaeda and the Islamic State in Iraq did before it.

Moreover, the situation in eastern Syria has become even more muddied after the US partial withdrawal, as the Kurds and various Tribal factions now seek a new accommodation with the Syrian regime and Russia, at the same time as they try to maintain their force and local governance structures, in part supported by the US's continued presence around the oilfields in Deir Ezzor.[60]

Rebel Groups: Rivalry and Hostility

Rebel groups associated with the Free Syrian Army have from early on been fighting Syrian government forces and their relations have continually been marked by war and enmity. When government forces retreated,[61] local councils, NGOs and rebel groups filled the void, providing basic

56 Hunaidi, 'ISIS has not been defeated'.
57 Becker, 'Terrorists and Target Governments Cooperate'.
58 Christophe Lafaye, *Habib Souaïdia, La sale Guerre.* (Paris: La Decouverte, 2001).
59 Lasse Lykke Rørbæk. 'Religion, Political Power, and the 'Sectarian Surge': Middle Eastern Identity Politics in Comparative Perspective'. *Studies in Ethnicity and Nationalism,* 19:1 (2019).
60 ICG, 'Lessons from the Syrian State's return to the South'.
61 This today in effect amount to the larger Idlib province and the Turkish dominated Euphrates Shield area (in addition to the North-East dominated by the SDF / Kurdish forces / tribes as described above). But back in 2014 different rebel groups held more than 70 per cent of Syrian territory.

healthcare services, water, electricity and roads, undertaking police work, manning check points, guarding warehouses and border crossing to extract resources and tax goods and services.[62] At the same time governance and service provision have always been inconsistent and extremely difficult as opposition-held areas have been marked by rebel infighting, heavily bombarded and besieged by the Assad government and Russia.

Armed opposition groups have, however, 'performed' much like a state in Timothy Mitchell's sense of the term.[63] To Mitchell, stateness rather than an already given state of being, that can be defined a priori, is a continuous social process that needs to be performed and enacted. This performance can for instance be the delivery of goods and infrastructure, emulating the 'tasks' traditionally associated with Syrian statehood before the war and filling the governance vacuum that the Assad government retreat created. The opposition groups' efforts to 'perform the state' has even implied setting up bakeries, subsidizing bread or distributing food baskets to build up local support and control. As Martinez and Eng argue, such rebel practices of 'state performance' are just as important as the military side in a war, and partially explain why the Assad regime deliberately bombed queues at bakeries or health clinics in opposition held areas, seeking to undermine rebels' ability to act and perform like a state.[64] Even jihadi rebel groups such as Hayat Tahrir al Sham (HTS) (previously Jabhat al Nusra) have also increasingly turned their attention to Islamist governance in addition to fighting the regime and rival rebel groups. HTS is running the so-called Salvation Government in Idlib, with ministers and countless administrative bodies and courts. It tolls commercial traffic from the border crossing at Bab al Hawa and the main highways inside the province, paves roads and controls an electricity utility.[65]

While rebel groups both fight and govern in order to hold on to territory and, ultimately, to overthrow the Assad government, none of the armed groups, from Ahrar al Sham, to Free Syrian Army or even the Kurdish groups, have over the course of the war contested the territorial boundaries or the international sovereignty of the Syrian state as a whole. They are not

62 Also, it should be noted by looting and taking their cut of trade and humanitarian aid, see e.g. ICG, 'Lessons from the Syrian State's return to the South', 14.

63 Timothy Mitchell, 'Society, Economy and the State Effect' in *State/Culture: State-Formation after the Cultural Turn*, ed. George Steinmetz (Ithaca and London: Cornell University Press, 1999).

64 José Ciro Martinez and Brent Eng. 'Struggling to Perform the State: The Politics of Bread in the Syrian Civil War'. *International Political Sociology*, 11 (2017), 131.

65 ICG, 'Steadying the new Status Quo', 9-20. However, it should be stressed HTS is also engaged in a range of mafia like activities and continues to carry out insurgency and suicide attacks.

secessionist rebel groups that seek their own state but can perhaps best be described as insurgents that seek to change the governing structures of Syria, while operating within its existing international territorial boundaries. Indeed, many of the opposition's administrative laws have deliberately been taken from Syrian state law 107, with some modifications, in order to be integrated more easily into what opposition actors at the time hoped would be future Syrian state law.[66] In this sense, even die-hard rebel groups do not only subvert or undermine state institutions but may also work to reproduce state practices and laws, paradoxically instilling a form of state continuity in their practice.

Though rebel groups share the goal of toppling the Assad government, they have from early on been engaged in intense infighting. Groups associated with the Free Syrian Army (FSA), and some of the more Islamist groups such as Jaish al Islam, Ahrar al Sham, or jihadi terrorist groups Jabhat al Nusra, Hurras al Din and the Islamic State, have fought each other, merged into new groups, or dissolved altogether. The FSA has often found itself in a three-front war, battling the Assad government and allied forces, the Islamic State and rival rebel groups concurrently. In the last rebel-held stronghold of Idlib, Jahbra al Nusra's successor, HTS, has taken over much of the province, while new fights have erupted between the Turkish backed National Liberation Front (NLF) and HTS and with regime forces in the southern part of the province. This increasingly scattered and fragmented rebel landscape has also had profound consequence for governance and service provisions as rival groups have set up parallel structures and/or competed with civilian bodies and NGO's.[67] Presently, armed actors in Idlib are reportedly fragmenting to such an extent that even in the same street one house may be controlled by a particular armed faction, while the next may be in the hands of another.[68]

External support for rebel groups came initially from a wide array of competing external actors such as Qatar, Saudi-Arabia and Turkey, all of whom sought to dominate the Syrian rebellion and draw it closer to their own agenda.[69] This inevitably contributed to the fragmentation of and infighting between the armed opposition groups. However, as the war evolved Gulf countries and the US have increasingly taken a back seat.

66 According to interview with Syrian activist in Idlib working closely with the Local Councils there, May 2018.
67 See also Malmvig, 'Mosaics of Power'.
68 ICG, 'Steadying the new Status Quo'.
69 Samer Abboud. *Syria.* (Cambridge: Polity Press, 2016); Phillips, 'Sectarianism as Plan B'.

It is now primarily Turkey who runs and supports armed groups in the Northwest. In the Euphrates Shield area where Turkey militarily intervened in 2016, Turkish dominance is particularly visible. This small pocket of land may best be described as a Turkish protectorate, as Ankara provides salaries for local councils, the police and the so-called National Army who all are trained in Turkey. Ankara also implements all major infrastructure, water, electricity and highways, just as identity cards are linked to the Turkish civil register. Turkish is also taught in the local Syrian schools.[70]

Though the Syrian state has been largely absent in rebel held areas, illegal trading with the Assad government through intermediaries and warlords in and out of the besieged areas of Ghouta and Douma have flourished, just as some civil servants and school teachers have continued to be paid by the Syrian government. Surprisingly, some of the very same people that had been on government payrolls were arrested when the regime retook Douma, Ghouta and the South in 2018.[71] In the South and around Damascus, rebels who were not killed or transported to Idlib, have 'reconciled' or surrendered through Russian security guarantees, and some have now enrolled into the Syrian Army's Fifth Corps sponsored by Russia. Allegedly they have had no other choice if they wanted to remain in their home area and/or provide for their families.[72] In Idlib, such forms of re-integration into the Syrian Army are however unlikely to happen. Here the rebel factions are dominated by jihadist groups and battle-hardened rebels who have arrived from other areas in Syria, unwilling or unable to compromise with the regime.[73] The situation in Idlib is thus a clear case of what Staniland's calls *guerilla disorder* and what many intuitively associate with civil war and state-armed non-state relations. These are characterized by outright hostility, no cooperation and unrestrained protracted violence, which may carry on for years.

70 Gregory Waters. 'Between Ankara and Damascus: The role of the Turkish State in north Aleppo'. Washington DC: Middle East Institute, June 20, 2019.
71 ICG, 'Lessons from the Syrian State's return to the South'.
72 ICG, 'Lessons from the Syrian State's return to the South', 5.
73 Haid, 'Reintegrating Syrian Militias'.

Conclusion: The not so Non-State and the Implications for our Understanding of the Syrian (State) Order

This chapter has described a fragmented and complex landscape of powerful armed non-state actors. Syria's many armed groups engage in multiple state-like tasks and often do so in close connection with and/or proximity to the Syrian state, blurring the line between what is state and non-state. The 'non-state' prefix may therefore in many ways be a misnomer, obscuring insights into how closely interwoven with state power, or state-like, non-state armed groups are.[74] Whether aligned with, or in opposition to, the Syrian government, armed groups perform a range of typical state tasks in terms of security, policing, extraction and infrastructure. Such state practices may be more or less state sanctioned, or state tolerated, as when armed Kurdish groups patrol border crossings, the Islamic State trades oil and electricity, or Hezbollah-supported militias pave roads or set up their own checkpoints. Even in rebel-held areas, the state has never been entirely absent, but has engaged in trade of vital resources or paid government salaries to teachers or intelligence officers.

Some of the practices of armed groups clearly challenge or subvert the authority of the Syrian state, as when militias who are formally aligned with the regime refuse to abide by a local ceasefire, deny the Syrian Army access or keep competing lists of males for military services. But, simultaneously, armed groups, also on the rebel-side, engage in practices that sustain, reproduce and uphold the state. The informal or de facto outsourcing of state tasks to armed groups has enabled the regime to survive, upholding its claim to statehood internationally, and has saved it vital resources for fighting and governing elsewhere.[75] Thus, the assumption that state and non-state armed actors necessarily compete in a form of zero-sum logic, where the latter erodes or takes something away from the state, is less helpful for those seeking to understand the very complex relations of overlapping and substituting networks of authority in Syria.

That having been said, this does not reversely imply that the Syrian state is intact, in control of its territory or that the government rules smoothly over, or through, the so-called loyalist or auxiliary armed non-state actors. Rather, Syria can be seen as a hybrid order of shifting nodes

74 See also Peter Baker. 'Heavy Pressure Led to Decision by Obama on Syrian Arms'. *The New York Times*, June 14, 2013; Lars Buur and Steffen Jensen. 'Introduction: Vigilantism and the policing of everyday life in South Africa'. *African Studies*, 63:2 (2004).

75 See also Leenders and Giustozzi, 'Outsourcing state violence'.

of governance, a space in which three school systems, border controls, currencies, and official documents intersect and compete, and where political authority is bargained and negotiated between the state and multiple armed non-state actors. Instead of resorting to concepts of state weakness or failure, Syria may therefore be approached as a fragmented political order of overlapping and entangled authorities, where multiple and shifting armed groups intersect, overlie and at times substitute the Syrian government, thereby blurring the boundary and distinctiveness between state and non-state. Finally, thinking beyond the modern notion of the territorially bounded state of monopolized violence, and the associated assumption that state and armed non state actors necessarily engage in a struggle over that monopoly, may perhaps also help us to appreciate the fallacies of the war propaganda on both sides: the Syrian state is neither regaining its centralized authority as the Assad government continuously assert, nor is it just a hollow shell, as some scholars would claim.

For European policymakers and practitioners, this also implies a necessary realization that the Syrian war(s) is not near its end. Most likely Syria will continue to be a patchwork of overlapping political orders for the foreseeable future, some characterized by outright hostility, others by forms of shared sovereignty and collusion, even ungoverned spaces, where extremist groups can regroup and resurrect. Tragically, the bells continue to toll for Syria.

References

Abboud, S. 'Hard Road ahead for the Syrian Exile Government'.' Beirut: Carnegie Middle East Center, November 15, 2013.

Abboud, S. *Syria*. Cambridge: Polity Press, 2016.

Almohamad, H. and A. Dittmann. 'Oil in Syria between Terrorism and Dictatorship'.' *Social Sciences* (2016), 1-16.

Baker, P. 'Heavy Pressure Led to Decision by Obama on Syrian Arms'.' *The New York Times*, June 14, 2013.

BBC. 'Syria unrest: Who are the Shabiha?' *BBC News*, May 29, 2012.

Becker, M. 'When terrorists and Target governments Cooperate the Case of Syria'.' *Perspectives on Terrorism* (2015), 95-103.

Bakonyi, J. and K. Stuvøy. 'Violence and Social Order beyond the State: Somalia and Angola'.' *Review of African Political Economy* (2005), 359–382.

Buur, L. and S. Jensen. 'Introduction: Vigilantism and the policing of everyday life in South Africa'.' *African Studies* (2004).

Davis, D. 'Non-State Armed Actors, New Imagined Communities and Shifting patterns of Sovereignty and Insecurity in the Modern World'.' *Contemporary Security Policy* (2009), 221-245.

Dukhan, H. *State and Tribes in Syria: Informal Alliances and Conflict Patterns.* London: Routledge, 2018.

Dukhan, H. 'Critical analysis of attempts to co-opt the tribes in Syria'. LSE blog. Last modified May 2, 2019. https://blogs.lse.ac.uk/crp/2019/05/02/critical-analysis-of-attempts-to-co-opt-the-tribes-in-syria/.

Enab, B. 'A year into the Settlement: the Regime's fragile security and control rule the landscape'. Enab Baladi. Last modified June 30, 2019. https://english.enabbaladi.net/archives/2019/06/a-year-into-the-settlement-the-regimes-fragile-control-and-security-chaos-rule-the-landscape/.

Eaton, T., C. Cheng, R. Mansour, P. Salisbury, J. Yasigi Jihad and L. Khatib. 'Conflict Economies in the Middle East and North Africa'. London: Chatham House, 2019.

Faucon, B. and A. Omran. 'Islamic State Steps Up Oil and Gas Sales to Assad Regime'. *The Wall Street Journal,* January 19, 2017.

Al-Ghadawi, A. 'New Configuration in the Syrian Regime's Security Structure'. London: Chatham House, November 2019.

Haid, H. 'Reintegrating Syrian Militias: Mechanisms, Actors and Shortfalls'. Beirut: Carnegie Middle East Center, December 12, 2018.

Haid, H. 'HTS's Offline Propaganda: Infrastructure, Engagement and Monopoly'. London: International Centre for the Study of Radicalisation, September 18, 2019.

Hinnebusch, R. *After the Arab uprisings: between democratization, counter-revolution and state failure.* London: Routledge, 2016.

Hughes, G. A. 'Syria and the Perils of proxy Warfare'. *Small Wars and Insurgencies* (2014), 522-538.

Hunaidi, S. 'ISIS has not been defeated. It's alive and well in Southern Syria'. *Foreign Policy,* April 3, 2019.

Hussein, A. '2018 Divisions defines Syria's tribes and Clans'. London: Chatham House, January 2018.

Heydemann, S. 'Beyond Fragility: Syria and the Challenges of reconstruction in Fierce States'. Washington DC: Brookings Institute, June 2018.

Hoffman, F. 'The Contemporary Spectrum of Conflict: Protracted, Gray Zone, Ambiguous, and Hybrid Modes of War'. Washington DC: The Heritage Foundation, 2016 available at https://s3.amazonaws.com/ims-2016/PDF/2016_Index_of_US_Military_Strength_ESSAYS_HOFFMAN.pdf.

International Crisis Group. 'Popular Protest in the Middle East and North Africa: the Syrian Regime's Slowmotion Genocide'. *Report 109,* Brussels: International Crisis Group, July 13, 2011.

International Crisis Group. 'Syria's Mutating Conflict'. *Briefing 128,* Brussels: International Crisis Group, August 1, 2012.

International Crisis Group. 'Avoiding a Free-for-all in Syria's North East'. *Briefing 66,* Brussels: International Crisis Group, December 21, 2018.

International Crisis Group. 'Lessons from the Syrian State's Return to the South'. *Report 196,* Brussels: International Crisis Group, February 25, 2019a.

International Crisis Group. 'Steadying the new Status Quo in Syria's North East'. *Briefing 72*, Brussel: International Crisis Group, November 27, 2019b.

Khaddour, K. 'Strength in Weakness: The Syrian Army's Accidental Resilience'. Beirut: Carnegie Middle East Center, March 16, 2016.

Kheder K. 'Local Wars and the Chance for Decentralized Peace in Syria'. Beirut: Carnegie Middle East Center, March 28, 2017.

Khateb, K. 'Hayat Tahrir al Sham in consultations on planned military and civil administrations in Idlib'. London: Chatham House, February 2019.

Khatib, L. and L. Sinjab. 'Syria's Transactional State: How the Conflict Changed the Syrian State's Execise of power'. London: Chatham House, October 2018.

Krause, K. and J. Milliken. 'The Challenge of Non-State Armed groups'. *Contemporary Security Policy*, 30:2 (2009), 202-220.

Leenders, R. and A. Giustozzi. 'Outsourcing state violence: The National Defence Force, 'stateness' and regime resilience in the Syrian war'. *Mediterranean Politics*, 24:2 (2017).

Lund, A. 'Syrian Jihadism'. Stockholm: Swedish Institute of International Affairs, September 14, 2012, available at http://www.sultan-alamer.com/wp-content/uploads/77409.pdf.

Lund, A. 'Blame Game over Syrians Stranded in the Desert'. Washington DC: The Century Foundation, June 18, 2018.

Malmvig, H. 'Coming in from the Cold: How we make take sectarianism seriously without playing to the tunes of regional power elites'. *Study 16*, Washington DC: Project on Middle East Political Science, September 17, 2015.

Malmvig, H. 'Mosaics of Power: Fragmentation of the Syrian State since 2011'. Copenhagen: Danish Institute for International Studies, 2018.

Martinez, J. C. and B.Eng. 'Struggling to Perform the State: The Politics of Bread in the Syrian Civil War'. *International Political Sociology* (2017), 130-147.

al Masri, Abdulrahman. 'Analysis: The Fifth Corps and the State of the Syrian Army'. *News Deeply*, January 11, 2017, available at https://www.newsdeeply.com/syria/articles/2017/01/11/analysis-the-fifth-corps-and-the-state-of-the-syrian-army.

Mehchy, Z. 'Back in Control, Syria's Regime tries to build its Legitimacy'. London: Chatham House, December 13, 2018.

Mitchell, T. 'Society, Economy and the State Effect'. *State/Culture: State-Formation after the Cultural Turn*, edited by George Steinmetz, Ithaca and London: Cornell University Press, 1999.

Özkizilcik, Ö. 'A way out for Russia and Turkey from Idlib's spiral of violence'. Washington DC: Middle East Institute, July 1, 2019.

Phillips, C. *The Battle for Syria: International Rivalry in the New Middle East*. New Haven: Yale University Press, 2016.

Phillips, C. 'Syria after IS'. *Orient*, 5 (2017). Available at http://www.orient-online.com/issues/ORIENT-IV/2017.

Phillips, C. 'Sectarianism as Plan B: Saudi-Iranian identity politics in the Syria Conflict'. *Saudi Arabia and Iran: The Struggle to Shape the Middle East*, edited by Simon Mabon, London: Foreign Policy Centre, 2018.

Polyakova, A. 'Russia Is a Great Power Once Again: Putin's Syria intervention saved Assad. But is he ready for what comes next?' *The Atlantic*, February 26, 2018.

Rauta, V. 'Towards a typology of non-state actors in "hybrid warfare": proxy, auxiliary, surrogate and affiliated forces'. *Cambridge Review of International Affairs*, September 9, 2019, available at https://doi.org/10.1080/09557571.2019.1656600.

Ruggie, J. G. 'Territoriality and Beyond: Problematizing Modernity in International Relations'. *International Organization*, 47:1 (1993), 139-174.

Rørbæk, L.L. 'Religion, Political Power, and the 'Sectarian Surge': Middle Eastern Identity Politics in Comparative Perspective'. *Studies in Ethnicity and Nationalism*, April 12, 2019, available at https://doi.org/10.1111/sena.12292.

Schlichte, K. *In the Shadow of Violence: The Politics of Armed Groups*. Frankfurt & New York, NY: Campus Verlag, 2009.

Syrian Law Journal. 'The New Urban Renewal Law in Syria', May 14, 2018, available at http://www.syria.law/index.php/new-urban-renewal-law-syria/.

Sottimano, A. 'Building authoritarian 'legitimacy': domestic compliance and international standing of Bashar al-Asad's Syria'. *Global Discourse*, September 9, 2016. https://doi.org/10.1080/23269995.2016.1212513.

Staniland, P. 'States, Insurgents, and Wartime Political Orders'. *Perspectives on Politics*, 10:2 (2012), 243-264.

Staniland, P. 'Militias, Ideology, and the State'. *Journal of Conflict Resolution*, 59:5 (2015).

Stubbs, J. 'Four-fifths of Russia's Syria strikes don't target Islamic State: Reuters analysis'. *Reuters*, October 21, 2015.

al-Tamimi, A. J. 'Administrative Decisions on Local Defence Forces Personnel: Translation & Analysis'. Last modified May 3, 2017, available at http://www.aymennjawad.org/2017/05/administrative-decisions-on-local-defence-forces.

al-Tamimi, A. J. 'The Story of al-Hajj Azra'il: From Nubl to Deir az-Zor'. Last modified March 16, 2018, available at http://www.aymennjawad.org/2018/03/the-storyof-al-hajj-azrail-from-nubl-to-deir-az.

Tejel, J. *Syrian Kurds: History, Politics and Society*. London: Routledge, 2019.

Van Veen, E. 'Creating a new Syria: Property, Dispossession, and Regime survival'. *Syria Comment*, April 15, 2018, available at https://www.joshualandis.com/blog/creating-a-new-syria-property-dispossession-and-regime-survival-by-erwin-van-veen/.

Vignal, L. 'The changing borders and borderlands of Syria in a time of conflict'. *International Affairs*, 93:4 (2017), 809-827.

Vinci, A. 'Anarchy, failed States and Armed Groups: Reconsidering Conventional Analysis'. *International Studies Quarterly* (2018), 295-314.

Waters, G. 'Between Ankara and Damascus: The role of the Turkish State in north Aleppo'. Washington DC: Middle East Institute, June 20, 2019.

Yahya, M. 'What Will It Take for Syrian Refugees to Return Home?' *Foreign Affairs*, May 28, 2018.

Zaman, T. *Islamic Traditions of Refuge in the Crises of Iraq and Syria*. London: Palgrave Macmillan, 2016.

Zohar, E. 'A New Typology of Contemporary Armed Non-State Actors: Interpreting the Diversity Studies in Conflict and Terrorism'. *Studies in Conflict and Terrorism* (2015), 423-450.

4

VIOLENT NON-STATE ACTORS WRITE THEIR OWN LAWS OF WAR

Jack Watling

Introduction

In the post-Cold War era, state failure and civil wars have proliferated, leaving large areas of Africa, the Middle East and Central Asia afflicted by Violent Non-State Actors (VNSAs). Concurrently, developed states have reduced the size of their armed forces[1] and sought to gain force multiplying capacity in the form of Private Military Companies (PMCs).[2] As the UK's Development, Concepts and Doctrine Centre concluded in 2014, 'by 2035, some non-state actors – multinational corporations, non-governmental organisations and city authorities – are likely to be more influential than they are today, having a greater impact on world affairs'.[3] At the same time, there has been a putative diminution of state sovereignty, caught between the growing power of supranational authorities,[4] and the tendency for

1 Magnus Nordenman, 'The Incredible, Shrinking Modern Military', *The Atlantic* (12 November 2012).

2 Peter Singer, *Corporate Warriors: The Rise of the Privatized Military Industry* (London: Cornell University Press, 2008); Sean McFate, *The Modern Mercenary: Private Armies and What They Mean for World Order* (Oxford: Oxford University Press, 2014).

3 DCDC, *Strategic Trends Programme: Future Operating Environment 2035* (London: Ministry of Defence, 2013), 11.

4 Martin Loughlin, 'The Erosion of Sovereignty', *Netherlands Journal of Legal Philosophy*, 2016:2, 2017, 57-81.

non-state actors to act transnationally.[5] This has changed the character of many civil wars from conflicts between parties over control of the state to struggles between competing non-state actors with a spectrum of objectives.[6] These conflicts are, correspondingly, transnational and, exacerbated by changes to international norms during the unipolar post-Cold War era, external intervention has been a common feature. As the world re-enters an era of great power competition, the indirect involvement of multiple state actors has enabled the proliferation of high-end weaponry,[7] accelerated by state failure, with the consequent distribution of state arsenals.[8] As General Richard Barrons notes, 'proxy warfare is the most successful kind of political war being waged of our generation'[9] and, for that very reason, is likely to continue to spread. States therefore have not only lost a monopoly on legitimate violence, but also on the tools of high-end warfare.[10]

These trends have important implications for how the regulation of war is approached. The foundations of contemporary international law rose from the codification in treaties of customs and norms that existed between states. The period from 1860-1960 arguably encompasses the zenith of the Westphalian international order and saw the codification of interstate relations. During the 1960s, the spread of mercenary warfare in Africa,[11] global terrorism[12] and a series of civil wars led to treaties and case

---

5 The movement of corporate earnings between states is both shaped by, and shapes state policy, see Kimberly A. Clausing, 'Multinational Firm Tax Avoidance and Tax Policy', *National Tax Journal*, 62:4, 2009, 703-725; comparably, the as Al Qaeda had moved to a transnational franchise model, so have those combatting it expanded the borders of their armed struggle to not just a global war on terror, but a war that is agnostic of state borders, see Brian Fishman, 'Using the Mistakes of Al Qaeda's Franchises to Undermine Its Strategies', *The Annals of the American Academy of Political and Social Science*, 618, 2008, 46-54.

6 Consider for example Libya, Yemen, Mali, or Syria, none of which can be said to be bipolar conflicts.

7 Such as the provision of ballistic missiles to non-state actors by Iran, see Robert Einhorn and Vann van Diepen, *Constraining Iran's Missile Capabilities* (Washington DC: Brookings Institution, 2019).

8 Nicholas Marsh, 'Brothers Came Back with Weapons: The Effects of Arms Proliferation from Libya', *PRISM*, 6:4, 2017, 79–96.

9 Richard Barrons, remarks delivered at the conference 'The Warfighter in the Twenty-First Century', *RUSI*, London, 23 November 2017

10 From ATGMs, see Jacub Janovsky 'Seven Years of War – Documenting Syrian Rebel Use of Anti-Tank Guided Missiles', *Bellingcat* (4 May 2018), to MANPADS, see Farzin Nadimi and Michael Knights, 'Iran's Support to Houthi Air Defenses in Yemen', *Policywatch*, no. 2953 (Washington DC: The Washington Institute for Near East Policy, 4 April 2018), and Armoured Vehicles, see Jack Watling, 'After Effect: The Future Looks Bleak for Iraq Post-Mosul Assault', *Jane's Intelligence Review* (December 2016), 30–33.

11 Exemplified in the Congo, see Mike Hoare, *Congo Mercenary* (London: Robert Hale, 1967).

12 Paul Chamberlin, *The Global Offensive: The United States, the Palestine Liberation Organization, and the Making of the Post-Cold War Order* (Oxford: Oxford University Press, 2015).

law pertaining to non-state actors. While non-state actors have increasingly become subjects of international law, they are not signatories to it and are only able to influence treaties by lobbying states. Thus, the agreements pertaining to non-state actors depend upon state enforcement. In many contemporary conflicts, states work in coalition and mutual dependency with non-state actors and often lack disproportionately greater military capabilities, making state enforcement of norms even more difficult. In this context, it is vital to consider how International law and, specifically, how International Humanitarian Law (IHL) functions, whether these frameworks are satisfactory and whether it is necessary to pursue the development of new norms in consultation with non-state actors.

This chapter is not primarily about the law; rather it is about how international law is perceived by one particular group of violent non-state actors: the Popular Mobilisation Forces (PMF) in Iraq. The chapter provides an ethnographic study of how a constellation of VNSAs defined the ethical and legal boundaries of their conduct on the battlefield and how they perceived the legal proclamations of Coalition forces operating in Iraq. The PMF did not claim to adhere to international legal frameworks. However, they did have internal codes of practice. The contention of this chapter is that, as VNSAs become increasingly prevalent on the battlefield, state forces may need to evaluate even more carefully their approach to promoting compliance with international law among partners and be increasingly open to explaining their actions in terms that also make sense to the groups they share the battlefield with – one audience that has often been overlooked. While many PMF practices were violations of international law, there were also significant areas of convergence, and, if states are more prepared to engage with VNSAs, it seems reasonable to suppose that norms of conduct could be agreed that would improve VNSA conduct in the field and avoid misunderstandings of states' rules of engagement. The chapter is based on fieldwork conducted in 2016, 2017 and 2019, which included extensive interviews with commanders, members and supporters of several VNSAs active in the conflict against Daesh, collectively described as the PMF. The author observed operations and conducted interviews with Iraqi politicians, government officials and military officers, along with officers, diplomats and officials from the Coalition to Defeat Daesh operating under the Combined Joint Task Force Operation Inherent Resolve.[13] The chapter

13 Owing to the vulnerability of some sources, and the sensitivity of issues surrounding the PMF in Iraq, some names of interviewees have been withheld.

is divided into four parts. The first will outline how the PMF emerged and the difficulties of trying to classify their status under international law. The second will detail the moral debates internal to the PMF over the rules governing their operations. The third will consider the perceived legitimacy of Coalition and PMF behaviour in Iraq. The chapter will conclude with policy recommendations for states seeking to maintain relevant policies and permissions to be effective in a similar future operating environment.

Iraq's Popular Mobilization Forces

In June 2014 Iraqi forces defending Mosul collapsed, leading to the city's capture by Daesh in just six days. The Iraqi Army, hollowed out by corruption,[14] and reviled for its sectarianism,[15] disintegrated, leaving its equipment to the militants. On Friday 13 June, Grand Ayatollah Ali al Sistani, the most senior cleric in Shia Islam, issued a Fatwa declaring that 'the defence of Iraq by our sons in the armed forces and the other security services is a sacred defence'.[16] He also made clear that the fight was a national effort against an external enemy, for 'the terrorists…are targeting all Iraqis in every area, so repelling them and fighting them is the responsibility of everyone, not just one sect or group to the exclusion of others'.[17] He called upon Iraqis to mobilise and they responded in the tens of thousands, far more than the security forces – barely able to track their own units – could absorb. Instead, these volunteers joined groups like the Badr Organization, Asaib Ahl al Haq, Khateib Hezbollah or Saraya al Salam, or formed new ones, including the Abbas Combat Division and Ali Al Akbar Brigade. Many did not formally join units, but simply drove to the front or to key locations like Samara, and took up arms, slowly gravitating towards more organised groups that could provide food, ammunition and logistics.

The result was a coalition of over 40 units with an unknown number of members. Some were established VNSAs like Khateib Hezbollah and Asaib Ahl al Haq. Some were political militias, associated with Iraqi MPs. Some

14 Dominic Evans, 'Iraq says it found 50,000 'ghost soldiers' on payroll', *Reuters* (1 December 2014).
15 Anna Louise Strachan, 'Factors behind the fall of Mosul to ISIL (Daesh) in 2014', *K4D* (17 January 2014): https://assets.publishing.service.gov.uk/media/59808750e5274a170700002c/K4D_HDR_Factors_behind_the_fall_of_Mosul_in_2014.pdf.
16 Abdul Mehdi Al-Karbala'i, addressed worshippers at the Imam Hussein Shrine in Karbala on 13 June 2014: http://www.sistani.org/arabic/archive/24918/.
17 Abdul Mehdi.

aligned with Badr were commanded by Federal Police officers.[18] Since Badr also controlled the Ministry of Interior, these officers were not forced to resign from government service and were often simultaneously policemen and militia commanders. Ali Al Akhbar Brigade was commanded by an officer in Iraqi Military Intelligence but answered to the Imam Hussein Shrine. All PMF units were theoretically to be coordinated by the Iraqi National Security Advisor through an entity known as the PMF Commission and made up of PMF commanders. Some openly declared that they served Ayatollah Khamenei, the Supreme Leader of Iran.[19] Others shunned Iranian advisers.[20] All ended up with a great deal of foreign equipment, from American Humvees and M109 self-propelled howitzers, to Iranian heavy machine guns and recoilless rifles and even T55 Main Battle Tanks taken from Iraqi arsenals.[21]

As the conflict went on, the administration of the PMF evolved. Once operations moved from defence to offence, ad hoc assembly became neither relevant nor viable. Instead, the numbers of PMF fighters declined to fit the logistics available. Offensive operations were more complex and lent themselves to having permanent, trained personnel. PMF units shifted from mass member organisations to battalions of around 500-800 men, formed into brigades of between 3,000 and 6,000.[22] Personnel were paid salaries and additional personnel were given basic training before being rotated into a reserve. Collectively the PMF shrank to a combat force of some 70,000 troops by mid 2016. The challenges of coordinating offensive operations had also required greater integration of PMF units into Iraqi military planning. By November 2016, partly in response to pressure from the international community to establish greater control over militias, the PMF were collectively made an official component of the Iraq security forces.[23] The force was assigned brigade numbers,[24] with state salaries restricted

---

18 The most notable being Abu Dergham al-Maturi who commanded 5 Badr Brigade while simultaneously holding office in the Federal Police.

19 Renad Mansour and Faleh al Jabar, 'The Popular Mobilization Forces and Iraq's Future', *Carnegie Middle East Center* (28 April 2017).

20 Sheikh Maytham al Zaydi, commanding the Abbas Combat Division, requested that Iranian assistance be routed through the Iraqi Ministry of Defence, as he explained to me in interview in September 2016.

21 Equipment observed on operations by the author in 2016 and 2017.

22 As explained in interview by Ahmad al Assadi in September 2019.

23 Saif Hameed, 'Iraqi parliament passes contested law on Shi'ite paramilitaries', *Reuters* (26 November 2016).

24 An index of these brigade numbers and their affiliations was compiled by Aymenn Jawad al-Tamimi, 'Hashd Brigade Numbers Index' (31 October 2017, updated 30 September 2019): http://www.aymennjawad.org/2017/10/hashd-brigade-numbers-index.

to 110,000 personnel. The PMF Commission managed the distribution of salaries. This was not done according to the number of personnel per unit but was instead doled out to PMF commanders to enable the commission to hold leverage over them.

Bringing the PMF legally under command of the Prime Minister did not in fact place PMF units completely under Iraqi control. Coordination was consultative. Power increasingly shifted to the PMF's commanders. Whereas the Iraqi Army had disgraced itself in 2014, the Iraqi public largely saw the PMF as ordinary Iraqis, volunteering to serve.[25] Following the recapture of Mosul, PMF commanders formed a political movement called the Fatah Coalition, which became the second largest bloc in Iraq's parliament in the elections of April 2018, winning a significant proportion of votes in Sunni as well as Shia areas.[26] Even accounting for low turnout, intimidation, ballot stuffing and other issues, it would be hard to argue that the PMF was less popular than the Iraqi government. Today the force remains a coalition of armed political groups, operating with the legal status of a component of the Iraqi armed forces.[27]

The designation of the PMF as clearly part of the Iraqi military remains problematic. While they are legally part of the state, many PMF groups pursue independent foreign policies. While the Iraqi Government says that it does not intend to interfere in the civil war in Syria,[28] PMF units have been sent to fight in Syria.[29] They have also engaged in other aggressive activities – such as conducting a strike on a Saudi pipeline in May 2019[30] – that are contrary to the stated aims and practice of Iraqi foreign policy. This is not a case of Iraq conducting a deniable foreign policy. Baghdad has, for some time, been pursuing closer ties with Riyadh.[31] Similarly, following the

25 Renad Mansour, 'The Popularity of the Hashd in Iraq', *Carnegie Middle East Centre* (1 February 2016).
26 Renad Mansour, 'More Than Militias: Iraq's Popular Mobilization Forces Are Here to Stay', *War on the Rocks* (3 April 2018).
27 Raya Jalabi, 'Iraq's Shi'ite militias formally inducted into security forces', *Reuters* (8 March 2018).
28 Ibrahim al Jafaari, speaking at the Royal United Services Institute, 14 September 2016: https://rusi.org/event/conversation-dr-ibrahim-al-jaafari-foreign-minister-republic-iraq.
29 Phillip Smith, 'From Karbala to Sayyida Zaynab: Iraqi Fighters in Syria's Shi`a Militias', *CTC Sentinel*, 6:8, 2013, 28-31.
30 Isabel Coles and Dion Nissenbaum, 'U.S.: Saudi Pipeline Attacks Originated From Iraq', *The Wall Street Journal* (28 June 2019); more specifically the attack was carried out by Khateib Hezbollah.
31 'Saudi Arabia: Back to Baghdad', *International Crisis Group* (22 May 2018): https://www.crisisgroup.org/middle-east-north-africa/gulf-and-arabian-peninsula/iraq/186-saudi-arabia-back-baghdad.

assassination of Qasem Suleimani, Khateib Hezbollah threatened, and has subsequently carried out, attacks on Iraqi military installations, targeting US personnel, that killed members of the Iraqi security forces.[32] This is despite the fact that Iraqi policy is currently to retain foreign trainers, although the continuation of that policy is a matter of debate.[33] The Iraqi government is both unwilling and unable to prevent PMF units pursuing independent policies. In interview, a senior Iraqi official, technically responsible for overseeing the PMF, turned his palms skywards in a gesture of helplessness when asked about these activities. Iraqi authorities cannot credibly hold these groups accountable, not least because one of the largest PMF groups is extensively embedded throughout the Ministry of Interior. In other words, they are not unaccountable due to a lack of state engagement but rather because of overt state engagement. Since they are not operating 'under the effective control' of the Iraqi government, their status is opaque. Some states have certainly treated them as an independent non-state actor. The US, for example, defines Khateib Hezbollah and Asaib Ahl a Haq as foreign terrorist organizations.[34] Despite both groups being a part of the Iraqi Security Forces under Iraqi law, the US government does not list the Iraqi government as a state sponsor of terrorism.[35]

Given this more complex situation, where groups like the PMF 'sit' in International law can be more difficult to assess. Consider, for instance, a theoretical operation by a PMF unit against Jabhat al Nusra in Syria in 2016. Approached from the perspective of Public International Law, lacking the legal capacity to act, the group is not permitted to use force. Ius ad bellum – international law pertaining to the use of force – reserves this right to states alone, within strict parameters. As Christian Henderson has observed, 'while the law on the use of force against non-state actors has...gone through a process of development over the past 20 years, the law governing the use of force by non-state actors has not gone through a similar development'.[36] In other words, while the Iraqi state could,

---

32 'UK soldier and two Americans killed in rocket attack in Iraq', *BBC* (12 March 2020).
33 Department of State, 'Joint Statement on the U.S.-Iraq Strategic Dialogue', 11 June 2020: https://www.state.gov/joint-statement-on-the-u-s-iraq-strategic-dialogue/.
34 Foreign Terrorist Organizations, Bureau of Counterterrorism, US State Department, https://www.state.gov/foreign-terrorist-organizations/.
35 State Sponsors of Terrorism, Bureau of Counterterrorism, US State Department, https://www.state.gov/state-sponsors-of-terrorism/.
36 Christian Henderson, 'Non-State Actors and the Use of Force', Math Noortmann, August Reinisch and Cedric Ryngaert (Eds.), *Non-State Actors in International Law* (Oxford: Hart Publishing, 2015), 77.

theoretically, use force in Syria, claiming self-defence (or by reference to a United Nations Security Council Resolution) using PMF forces, the PMF could not act independently. However, from the PMF unit's point of view, it was mobilised to defend the shrines and holy places of the Shia *Ummah*, which existed in Syria as well as Iraq. From this perspective, the group not only conceived of itself as a legitimate actor, but one that was operating within the territory it was legitimately defending.

Further difficulties apply to VNSA behaviour on the battlefield.[37] In the Battle for Mosul, for instance, the forces opposing Islamic State numbered approximately 90,000 troops. Of these, around a third comprised Iraqi army and police, a third were Kurdish Peshmerga and a third were from the PMF. Although answerable to the Kurdish Regional Government (KRG), Peshmerga units are best described as party affiliated militia and, in different contexts, straddle the divide between state and non-state actors.[38] On balance, the Iraqi state commanded the majority of forces, although, for much of the operation, control was a matter of negotiation. The Iraqi military, for instance, was unable to prevent the PMF participating in the assault on Tal Affar[39] and eventually made a virtue of their successes, despite not having directed the operation. In this context, the state lacked the power to compel the PMF to comply with IHL and building compliance would by necessity have to be voluntary. Coalition forces however were reluctant to build direct connections to PMF units for a number of reasons. Partly it reflected the US designation of some PMF units as terrorists and mistrust between elements of the PMF and the Coalition. However, legal concerns were another reason for the lack of engagement. If PMF units were advancing under the protection of Coalition airpower, executing operations that were derived from US and UK planning and employing techniques under the oversight of embedded special forces, the Coalition risked incurring obligations under the Articles governing the Responsibility

37 The actions of such groups have legal implications, not only for the VNSAs themselves who have obligations under IHL, but, potentially, for states with whom they are associated. If attribution can be established (and the nature of this varies considerably) states may find themselves in breach of international law – in connection with the acts of VNSAs for which they may be considered responsible or in connection with their failure to meet obligations of 'conduct' perhaps relating to the regulation of VNSA actions.

38 Mario Fumerton and Wladimir van Wilgenburg, 'Kurdistan's Political Armies: The Challenge of Unifying the Peshmerga Forces', *Carnegie Endowment for International Peace* (16 December 2015).

39 Ahmad Majidyar, 'P.M.F. Forces Reportedly Barred from Participating in Tal Afar Operation', *Middle East Institute* (10 August 2017): https://www.mei.edu/publications/pmf-forces-reportedly-barred-participating-tal-afar-operation.

of States for Internationally Wrongful Acts (ARSIWA), which state that '[t]he conduct of a person or group of persons shall be considered an act of a State under international law if the person or group of persons is in fact acting on the instructions of, or under the direction or control of, that State in carrying out the conduct'.[40] Given limited tools to shape PMF behaviour the Coalition kept its distance. Support was still necessary however and so Special Forces teams would embed with Iraqi units that would co-locate with PMF units to coordinate airstrikes.[41] The result was that the PMF were enabled, but the Coalition did not hold the leverage to shape their behaviour or enter into a dialogue. At the very least, changing this dynamic is likely to require the capacity to engage with VNSAs in meaningful discourse.

In Sacred Defence

Engaging in meaningful dialogue with VNSAs must be premised upon a degree of mutual respect. To be effective or even persuasive, state interlocutors must understand the moral and, in some more limited instances, legal frameworks within which VNSAs operate. Having considered the more complex nature of the PMF, we now turn to the issue of how the PMF itself described their own approach to both the law and the ethics of war.

The starting point for members of the PMF was the Fatwa of Grand Ayatollah Sistani.[42] Sistani is the most prominent jurist in Shia Islam, but he has no monopoly on juristic authority. Several PMF Groups have declared themselves followers of Ayatollah Khamenei and many members of the PMF follow other Ayatollahs. Sistani described Daesh as 'terrorists' and claimed that 'the way of these terrorists is against the spirit of Islam, it rejects co-existence with the other in peace and it adopts violence, bloodshed and provoking sectarian strife as a means of spreading its influence and power across different parts of Iraq and other countries'.[43] Some chose to interpret

40 ARSIWA, Chapter 2, Article 8: https://www.law.umich.edu/facultyhome/drwcasebook/Documents/Documents/International%20Law%20Commission%202001%20Draft%20Articles%20on%20State%20Responsibility.pdf.

41 Department of Defense Press Briefing by Col. Dorrian via teleconference from Baghdad, Iraq (22 February 2017): https://dod.defense.gov/News/Transcripts/Transcript-View/Article/1091450/department-of-defense-press-briefing-by-col-dorrian-via-teleconference-from-bag/.

42 Abdul Mehdi Al-Karbala'i, addressed worshippers at the Imam Hussein Shrine in Karbala on June 13, 2014: http://www.sistani.org/arabic/archive/24918/.

43 Abdul Mehdi.

Sistani's words as indicating that Daesh fighters were 'not true-Muslims' or were apostates who had abandoned their faith.[44] Others argued that Daesh fighters were Kharawij[45] – or violent extremists – and that the fight against them was a Jihad al Saif al Kharawij.[46] Iranian officials – notably Major General Qasem Suleimani in a letter endorsed by Ayatollah Khamenei[47] – argued that Daesh were 'takfiri criminals' who 'prevented Muslims from truly understanding the pure Islam of Muhammad' and thereby corrupted them to do their will. Iranian officials in particular alleged that Daesh was nurtured from outside the Muslim word, by the US and UK, a view that was far from uncommon within the PMF.[48] It is notable that Suleimani attempted to frame Iranian policy as consistent with Sistani's fatwa, writing that the defeat of Daesh was made possible by 'the wise leadership and guidance of his Excellency, Ayatollah Sistani, who mobilized all facilities to deal with this toxic storm'.[49] Nevertheless, PMF units that followed Ayatollah Sistani adhered to different codes of practice to those who followed Ayatollah Khamenei. Regardless of which cleric they adhered to, it must be recognised that, within Shi'ism, the Maraji – the senior clergy who are followed – are jurists, uniquely qualified as interpreters of Sharia, which they believe is God's law.[50] As the 1990 Cairo Declaration on Human Rights in Islam makes clear,[51] many Muslim majority states have objections to elements of the existing international legal framework. The Islamic Republic of Iran was a signatory to that declaration. However, it is vital to appreciate that, for the PMF, God's law transcends the law of nations; the interpreters of God's law are the Maraji and the onus is on states to demonstrate the consistency of their laws with Islamic principles, rather than the other way about.

---

44 Interview with a senior Iranian cleric, August 2017.

45 Mohamed bin Ali, 'Labelling IS Fighters: Khawarij, Not Jihadi-Salafis', *RSIS Commentary* (4 April 2018).

46 Ahmed al Dawoody, *The Islamic Law of War: Justifications and Regulations* (New York: Palgrave, 2011), 77.

47 Sayed Ali Khamenei, 'By eradicating US-backed ISIS you helped not only Muslims but also the mankind', *Khamenei.ir* (21 November 2017): http://english.khamenei.ir/news/5284/By-eradicating-US-backed-ISIS-you-helped-not-only-Muslims-but.

48 I was confronted as to why the US and UK had helped to create Daesh by a number of PMF officers over the course of 2016–2017.

49 Qasem Suleimani, 'Gen. Soleimani congratulates Ayatollah Khamenei and Muslims on ISIS termination', *Khamenei.ir* (21 November 2017): http://english.khamenei.ir/news/5283/Gen-Soleimani-congratulates-Ayatollah-Khamenei-and-Muslims-on.

50 P. Sadeghi and Jack Watling, 'The Constraints of Infallibility: The Pronouncements of Iran's Supreme Leader', *The RUSI Journal*, DOI: 10.1080/03071847.2020.1767510

51 Cairo Declaration on Human Rights in Islam, 5 August 1990.

That the PMF did not necessarily recognise international law did not mean that they were hostile to all its principles. In the autumn of 2016 and 2017, I interviewed PMF members who were on operations and PMF commanders in Karbala and Najaf, where they sought guidance from the Marjai'yah. There were ongoing and complex debates within units and between PMF commanders as to when it was appropriate to use certain munitions, how prisoners were to be classified and treated and the status of the local population. These discussions referenced an extensive body of religious case law, tradition and reasoning, without direct reference to the frameworks set out in international law. The language was different, but it is important to recognise that many concepts were convergent with wider principles in International Humanitarian Law. For instance, Sistani had instructed fighters to 'Make it your aim, your intention and your motivation to defend the sacred places of Iraq and its unity, to preserve the safety of it citizens, protect its holy places from damage, and to push the evil out of this country and away from its wounded people'.[52] One moral question this generated for commanders was when it was appropriate to use artillery during assaults to recapture urban terrain. On the one hand, the purpose of the campaign was to drive Daesh from Iraq, which could be accomplished with the fewest PMF casualties if Iraqi forces employed artillery. At the same time, those fires would destroy buildings and kill civilians, whose liberation was the declared purpose of the operation. Explaining the Marjaiy'ah's reasoning on this point, in September 2016, a senior cleric said that 'I do not advise commanders on what weapon to employ; this is for them to judge using their technical expertise. But the principle guiding their judgement should be to achieve a balance between their responsibility to defeat Daesh, to protect civilians and their property, and to preserve the lives of their soldiers'.[53] Where the line was drawn depended upon the PMF unit, the clerics they consulted for advice and the military situation that confronted them. For instance, several PMF units began to use 57mm anti-aircraft guns as artillery when advancing on urban areas.[54] This was because the gun was found to set off IEDs in a wide area, clearing a path for friendly forces. It was also noted by operators that the AA shell did less structural damage to buildings than impact-fused High Explosive (HE) rounds and therefore gave

52 Abdul Mehdi Al-Karbala'i, addressed worshippers at the Imam Hussein Shrine in Karbala on June 13, 2014: http://www.sistani.org/arabic/archive/24918/.
53 Interview with a senior ceric, Najaf, September 2016.
54 Some were mounted on T55 chassis, see Yri Lyamin, 'T-55 chassis fitted with S-60 57mm gun in Iraq', *The Hoplite* (26 October 2016).

civilians hiding in those buildings a greater probability of surviving. This is an example of how PMF units were prepared to change their operating practices in order to uphold principles that were convergent with those found in IHL. The debate outlined above is very similar to debates over the balance between proportionality and military necessity, which is central to the application of IHL. The process and language was fundamentally different, but did not reach drastically different practical conclusions.

There were also areas of significant divergence from IHL in some PMF practices. It is important, at the outset, to distinguish between actions in violation of IHL that arose from issues of training and combat friction and those that were deliberate acts of policy. This distinction does not change whether the act was a violation of IHL, but it does affect the extent to which the behaviour could have been reduced. To illustrate this distinction, consider the experiences of Mohsen, a rifleman in the Abbas Combat Division (ACD), interviewed in September 2016. Having mobilised a year before, Mohsen had seen fighting in Tikrit, Baiji and the Makhoul and Hamrin mountains. By this stage, 79 members of the ACD had been killed and between 230-250 had been wounded. Casualties would rise steeply in the months that followed. Three enemy tactics were responsible for several multiple casualty incidents. The first was Daesh's tendency to ram checkpoints with vehicle born improvised explosive devices (VBIEDs) at night. The ACD adopted a range of countermeasures. The most effective was to use their 57mm anti-aircraft guns, mounted on flatbed lorries, to fire at approaching cars. There was, PMF gunners conceded, little opportunity for a warning shot and so some civilian vehicles were mistakenly hit. It was effective, given that even a miss tended to set off the VBIEDs. The second occurred while taking prisoners, since a proportion of Daesh fighters would detonate themselves upon capture.[55] Mohsen recalled that many of the Daesh fighters were terrified, some would surrender, but a proportion would detonate grenades or suicide vests. He intimated that, as the conflict went on, his comrades took fewer risks. There were plenty of instances of prisoners being shot. The third threat was Daesh's leaving behind fighters in overrun positions, who would emerge later and carry out a suicidal rampage among Iraqi forces. In one case, Mohsen described taking a small village. After clearing the buildings and setting up a perimeter, one of the ACD Non Commissioned Officers (NCOs) got on a roof to hoist a flag so

55 As recently demonstrated by Abubakr al Baghdadi, see 'Abu Bakr al-Baghdadi: IS leader 'dead after US raid' in Syria', *BBC* (28 October 2019).

that friendly forces would know the position was captured. Four of his comrades stood back, watching. As the NCO emerged from the building, he was followed by a man dressed as a civilian, who raised a rifle, shot him in the back and then shot the four other members of the unit. The Daesh fighter was killed shortly after, but Mohsen explained that this, and incidents like it, left him deeply distrustful of the local population.

The confluence of these experiences – and what Mohsen had seen was not uncommon among Iraqi forces – led to widespread, though not systematic atrocities. When towns were retaken, Iraqi forces would advance to contact, then bring down heavy artillery fire. Some people were killed trying to surrender. Those that did surrender might be treated humanely or else beaten, killed or disappeared.[56] It did not take much for civilians to be suspected of being members of Daesh and, for them, the consequences were predictable. Further killings took place on the basis of denunciations, either from local residents or from Internally Displayed Peoples from the community concerned. Most of these killings were carried out on an ad hoc basis and to a great extent depended upon the decision of local commanders.[57] It is important to note that, in this respect, the PMF were not vastly dissimilar on occasion to the Iraqi Army. Ghaith Abdul-Ahad, reporting during the final days of the siege of Mosul, described in intimate detail 'an orgy of killing' carried out by PMF and Iraqi forces, which typified in style – though not in scale – what had taken place elsewhere.[58] Better tactics, techniques and procedures for prisoner management, sweeps, and perimeter defence could have reduced many of these excesses. They would not have been eliminated entirely, but greater or even more systematic engagement may have mitigated further and offered some hope for wider engagement with non-state actors.

At the same time, there were practices among the PMF that were much more deliberate violations of IHL – which indicates a need for caution about both the principle and practice of engagement. Asaib Ahl al Haq, for example, conducted sectarian massacres from early in the campaign, which were clearly pre-meditated.[59] Following the liberation of Iraqi towns, some PMF units practiced reprisals targeting the property of locals associated

56 '643 Iraqi Men Disappeared Two Years Ago: Where Are They?', *Amnesty International* (11 June 2017); updated 11 June 2018.
57 Nour Samaha, 'Iraq's 'Good Sunni'', *Foreign Policy* (16 November 2016).
58 Ghaith Abdul-Ahad, 'After the liberation of Mosul, an orgy of killing', *The Guardian* (21 November 2017).
59 Human Rights Watch, 'Iraq: Survivors Describe Mosque Massacre' (2 November 2014): https://www.hrw.org/news/2014/11/02/iraq-survivors-describe-mosque-massacre.

with Daesh fighters, often because of tribal connections rather than personal collaboration.[60] This collective punishment, often targeting civilians, is clearly a violation of IHL. Significantly, PMF commanders and clerics defended such actions in interview on a number of grounds. Justifications often included judgement regarding the collective guilt of specific tribes or communities or the view that it was not only Daesh's fighters but also the wider membership who posed a threat. Some commanders argued that Daesh atrocities justified responses in kind. Reference was made to Al Baqarah in the Quran, in which Allah directs that 'if then anyone transgresses the prohibition against you, transgress ye likewise against him. But fear Allah, and know that Allah is with those who restrain themselves'.[61] Given the excesses carried out by Daesh fighters, PMF fighters could argue that they were restrained by comparison, even while acting well beyond the limits imposed by IHL. It seems very unlikely that any amount of training or dialogue with Coalition forces would have shifted some of these beliefs. The challenge therefore is whether states feel that they are able to reach out to groups to discuss norms of conduct, and potentially offer training, that could reduce violations of IHL, while accepting that some violations will continue, without exposing personnel to legal jeopardy.

Legitimacy and Legitimation

The Universal Declaration of Human Rights begins by stating that 'the inherent dignity and…the equal and inalienable rights of all members of the human family is the foundation of freedom, justice and peace in the world'.[62] As Samuel Moyn makes clear in his history of the development of human rights and the international order, the modern system of international law was far from inevitable.[63] Nor is its continuation assured. There is a danger for those advocating adherence to international law that they assume what they are proposing is self-evidently good for all concerned. For states seeking to shape the behaviour of VNSAs, it is important to recognise that they cannot take their own legitimacy or the legitimacy of the

60 'Ruinous Aftermath: Militias Abuses Following Iraq's Recapture of Tikrit', *Human Rights Watch* (20 September 2015).
61 'Surah 2: Al Baqarah, 194', *The Holy Quran*, Translated by Abdullah Yusuf Ali (New Delhi: Kitab Bhavan, 2011), 34.
62 The Universal Declaration of Human Rights, Preamble, 10 December 1948.
63 Samuel Moyn, *The Last Utopia: Human Rights in History* (Cambridge MA: Harvard University Press, 2012).

international legal order for granted and must seek to explain compliance with international law in terms that also make sense to their audience. This dynamic is important in the contemporary operating environment. US and UK policy in Iraq sought to uphold the legitimacy of the Iraqi government, while avoiding the legitimation of VNSAs such as the PMF.[64] This position made the mistake of assuming that legitimacy in the eyes of Iraqis was the US and UK's to give.

Distinctions between groups that appear clear in Whitehall or Washington were less evident to many Iraqis. As Michael Knights, Hamdi Malik and Ayman Jawad al-Tamimi observe in the most detailed study of PMF formations to date, 'a consistent and strong positive sentiment toward the Hashd [PMF]…exists within Iraq'.[65] They note that 'nearly one million Iraqis directly depend on the Hashd for their livelihood. Hundreds of thousands more have an indirect but personal connection to the Hashd, and millions believe they owe the Hashd a debt of gratitude'.[66] This sentiment was not evenly felt between Shia and Sunni communities, but did not split directly or neatly along sectarian lines. Despite many Iraqis having negative sentiments towards some PMF groups, such as Khateib Hezbollah, there was widespread support for the PMF as an institution. Thus, Coalition refusal to work with the PMF undermined Coalition pronouncements that they were unequivocally backing the Iraqi war effort against Daesh and encouraged the perception that the Coalition was using the opportunity to play politics. On the other side of the lines the distinction between PMF and Iraqi military units was also unclear to many Iraqis. Throughout the campaign against Daesh, colleagues and I were sent pictures by residents of liberated towns claiming that they had been occupied by the PMF. The pictures usually showed flags with images of Shia Imams on them or soldiers with Shia insignia on their uniforms and vehicles. Sometimes these were pictures of PMF units. Often they were images of Iraqi soldiers who shared many religious beliefs with PMF fighters and used comparable symbols. Again therefore, Coalition claims to be supporting the Iraqi state, but not the PMF, made less sense to those on the ground, given the blurred perceptions of legitimacy.

---

64 As explained by CJTF-OIR Commander Lieutenant General Sean Macfarland, see 'Lt. Gen. Sean MacFarland: Building better Arab armed forces | LIVE STREAM', *American Enterprise Institute* (18 March 2019): 00:46:00: https://www.youtube.com/watch?v=EWHffWl0DUY.

65 Michael Knights, Hamdi Malik, and Ayman Jawad al-Tamimi, *Honored, Not Contained: The Future of Iraq's Popular Mobilization Forces* (Washington DC: The Washington Institute for Near East Policy, 2020), 148.

66 Michael Knights, 'Honored'.

Gaps between the narrative communicated by the Coalition and the experience of Iraqis also had serious implications for how Coalition compliance with IHL was perceived. The British government has claimed that it has evidence of only one civilian casualty in RAF airstrikes against Daesh.[67] This is likely to be technically true. Battle Damage Assessments on destroyed structures are unlikely to reveal the civilians that were taking shelter in the basement. Yet, this highlights how Defence has prioritised reassuring a domestic audience, rather than building legitimacy for its conduct on the ground. The number of strikes on buildings, combined with the tendency for civilians to conceal themselves during fighting, makes the claim of one civilian death untenable.[68] The British government has not interviewed residents, but, if it released a list of the sites hit, names of the dead would not be difficult to establish. Of course, this does not make the UK non-compliant with IHL. In many cases, hitting the building was consistent with the principle of military necessity, lower payload munitions were not always available or suitable and, if the RAF knew there were civilians present, they would not strike. The policy however has set an impossible standard, leading to charges of hypocrisy and reducing the legitimacy gap further between the agents of the state and the PMF. Furthermore, the portrayal of Western military operations in popular culture – and Defence's amplification of a narrative that inflates precision – means that many Iraqis do not see the civilian deaths as mistakes. Instead, it is often seen as deliberate. This is problematic because it leads Iraqis to the conclusion that Western pronouncements about international law are disingenuous and are, at best, a form of information warfare, enabling the West to portray its adversaries as non-compliant, while being dishonest about its own record. The perception that international law is a system designed to enable Western power, rather than a set of universal principles to establish norms between parties, undermines those calling for wider compliance with the underlying principles of IHL. It is deeply unfortunate that, despite a great deal of effort and resource within Operation Inherent Resolve (OIR) being devoted to ensuring the legality of targeting, the

67 Alexandra Heal, 'RAF says it harmed just one civilian out of 4,000 enemy fighters in Iraq and Syria', *The Guardian* (7 March 2014).

68 This was convincingly argued by Samuel Oakford, 'Credibility Gap: United Kingdom civilian harm assessments for the battles of Mosul and Raqqa', *Airwars* (September 2018): https://airwars.org/wp-content/uploads/2018/09/UK-Inquiry-into-Mosul-and-Raqqa-2018.pdf, 17 November 2019.

messaging around precision and collateral damage arguably harmed the credibility of UK forces.[69]

Given the contours of the information environment outlined above, Coalition forces cannot assume that there is much willingness to ascribe friendly intent to their policies, adding a further layer of complexity to attempts to increase engagement. This creates a challenge for credibly advocating compliance with IHL among partner forces, because the Coalition is often popularly perceived to lack the authority to make its advocacy have impact. In contrast, PMF units have retained political legitimacy and translated operational successes into political capital, despite violating IHL. This partly reflects the suppression of its critics but is also a consequence of the PMF being able to explain its conduct with reference to religious, cultural and political concepts that have greater resonance for Iraqis. As the International Committee of the Red Cross (ICRC) concluded in one of the most extensive studies of how best to shape armed group compliance with IHL, 'an exclusive focus on the law is not as effective at influencing behaviour as a combination of the law and the values underpinning it. Linking the law to local norms and values gives it greater traction'.[70] In short, if states wish to reinforce their legitimacy, and the perceived legitimacy of their behaviour, using the language of international law is likely far less effective than explaining actions in terms better understood by the audience. Without that legitimacy states will struggle to be listened to in advocating adherence to norms of conduct that they are bound by their treaty commitments to promote.

Implications for UK Forces

This chapter has sought to outline three points. First, the tendency for there to be a greater level of parity in power between VNSAs and state forces in a growing number of contemporary conflicts means that the model whereby states enforce international laws that apply to, without necessarily engaging with, VNSAs is increasingly not fit for purpose. Secondly, states should consider more fully and critically how and when

69 Detailed opinions polls are scarce on sensitive subjects like this. This conclusion is based on the frequency with which colleagues and I were asked by Iraqi civilians, officials, and officers – at various levels – to explain what they perceived to be duplicity in UK and US policy.

70 Fiona Terry and Brian McQuinn, *The Roots of Restraint in War* (Geneva: ICRC, December 2018), 65.

to engage VNSAs in dialogue about norms of conduct. Thirdly, states cannot assume sole or even primary legitimacy in these discussions but must explain the principles they are advocating in terms that make sense to VNSAs and the populations within which they operate without confusing wider international and domestic audiences. This is a difficult balancing act, as has been demonstrated above.

The British military has been restructuring itself to reflect the demands of the future operating environment. Across Defence, the Integrated Operating Concept seeks to clarify how the military can operate and use force beneath the threshold of full-scale armed conflict. The newly reformed 6 United Kingdom Division has the explicit purpose of conducting irregular warfare and is intended to compete using enablers, partner force capacity building and information operations, to project power onto an increasingly complex and convoluted battlefield. At the same time however, OIR in Iraq maintained a highly centralised, bureaucratic and process laden legal framework for approving any action. If OIR typifies the UK's expectation of how IHL is to be translated into operations, it is difficult to see how 6 UK Division and similar formations will have a sufficiently flexible license to operate effectively. Historically, Special Forces have led engagement with VNSAs. Beyond offering limited degrees of leverage over VNSA behaviour, it is unlikely that this policy is sustainable if the volume of engagement with VNSAs increases significantly. As it will need to be a task for conventional forces, and therefore come under greater legal scrutiny, it is vital that the UK works out appropriate permissions and procedures to protect personnel from legal jeopardy, while being able to be operationally effective. The price of making it even more difficult for intervening states to exert effective influence in the contemporary operating environment for fear of legal jeopardy will not be a reduction in violence, but instead the abandonment of the battlefield to VNSAs who will readily write their own laws of war.

References

'643 Iraqi Men Disappeared Two Years Ago: Where Are They?', *Amnesty International* (11 June 2017); updated 11 June 2018.

'Abu Bakr al-Baghdadi: IS leader 'dead after US raid' in Syria', *BBC* (28 October 2019).

Abdul-Ahad, G. 'After the liberation of Mosul, an orgy of killing', *The Guardian* (21 November 2017).

al-Dawoody, A. *The Islamic Law of War: Justifications and Regulations* (New York: Palgrave, 2011).

al-Karbala'i, A. M. Address to Worshippers at the Imam Hussein Shrine in Karbala on 13 June 2014: http://www.sistani.org/arabic/archive/24918/.

al-Jafaari, I. speaking at the Royal United Services Institute, 14 September 2016: https://rusi.org/event/conversation-dr-ibrahim-al-jaafari-foreign-minister-republic-iraq.

al-Tamimi, A. J., 'Hashd Brigade Numbers Index' (31 October 2017, updated 30 September 2019): http://www.aymennjawad.org/2017/10/hashd-brigade-numbers-index.

Barrons, R. remarks delivered at the conference 'The Warfighter in the Twenty-First Century', RUSI, London, 23 November 2017

bin Ali, M. 'Labelling IS Fighters: Khawarij, Not Jihadi-Salafis', *RSIS Commentary* (4 April 2018).

Cairo Declaration on Human Rights in Islam, 5 August 1990.

Chamberlin, P. *The Global Offensive: The United States, the Palestine Liberation Organization, and the Making of the Post-Cold War Order* (Oxford: Oxford University Press, 2015).

Clausing, K. 'Multinational Firm Tax Avoidance and Tax Policy', *National Tax Journal*, 62:4, 2009, 703-725.

Coles, I. and Nissenbaum, D., 'U.S.: Saudi Pipeline Attacks Originated From Iraq', *The Wall Street Journal* (28 June 2019).

DCDC, *Strategic Trends Programme: Future Operating Environment 2035* (London: Ministry of Defence, 2013).

Department of Defense Press Briefing by Col. Dorrian via teleconference from Baghdad, Iraq (22 February 2017): https://dod.defense.gov/News/Transcripts/Transcript-View/Article/1091450/department-of-defense-press-briefing-by-col-dorrian-via-teleconference-from-bag/.

Department of State, 'Joint Statement on the U.S.-Iraq Strategic Dialogue', 11 June 2020: https://www.state.gov/joint-statement-on-the-u-s-iraq-strategic-dialogue/.

Einhorn, R. and van Diepen, V. *Constraining Iran's Missile Capabilities* (Washington DC: Brookings Institution, 2019).

Evans, D. 'Iraq says it found 50,000 'ghost soldiers' on payroll', *Reuters* (1 December 2014).

Fishman, B. 'Using the Mistakes of Al Qaeda's Franchises to Undermine Its Strategies', *The Annals of the American Academy of Political and Social Science*, 618, 2008, 46-54.

Foreign Terrorist Organizations, Bureau of Counterterrorism, US State Department, https://www.state.gov/foreign-terrorist-organizations/.

Fumerton, M. and van Wilgenburg, W. 'Kurdistan's Political Armies: The Challenge of Unifying the Peshmerga Forces', *Carnegie Endowment for International Peace* (16 December 2015).

Hameed, S. 'Iraqi parliament passes contested law on Shi'ite paramilitaries', *Reuters* (26 November 2016).

Heal, A. 'RAF says it harmed just one civilian out of 4,000 enemy fighters in Iraq and Syria', *The Guardian* (7 March 2014).

Henderson, C. 'Non-State Actors and the Use of Force', Noortmann, M., Reinisch, A. and Ryngaert, C. (Eds.), *Non-State Actors in International Law* (Oxford: Hart Publishing, 2015).

Hoare, M. *Congo Mercenary* (London: Robert Hale, 1967).

Human Rights Watch, 'Iraq: Survivors Describe Mosque Massacre' (2 November 2014): https://www.hrw.org/news/2014/11/02/iraq-survivors-describe-mosque-massacre.

Jalabi, R. 'Iraq's Shi'ite militias formally inducted into security forces', *Reuters* (8 March 2018).

Janovsky, J. 'Seven Years of War – Documenting Syrian Rebel Use of Anti-Tank Guided Missiles', *Bellingcat* (4 May 2018).

Khamenei, A., 'By eradicating US-backed ISIS you helped not only Muslims but also the mankind', *Khamenei.ir* (21 November 2017): http://english.khamenei.ir/news/5284/By-eradicating-US-backed-ISIS-you-helped-not-only-Muslims-but.

Knights, M., Malik, H. and al-Tamimi, A. J. *Honored, Not Contained: The Future of Iraq's Popular Mobilization Forces* (Washington DC: The Washington Institute for Near East Policy, 2020).

Loughlin, M. 'The Erosion of Sovereignty', *Netherlands Journal of Legal Philosophy*, 2016:2, 2017, 57-81.

'Lt. Gen. Sean MacFarland: Building better Arab armed forces | LIVE STREAM', *American Enterprise Institute* (18 March 2019): 00:46:00: https://www.youtube.com/watch?v=EWHffWl0DUY.

Lyamin, Y. 'T-55 chassis fitted with S-60 57mm gun in Iraq', *The Hoplite* (26 October 2016).

Majidyar, A. 'P.M.F. Forces Reportedly Barred from Participating in Tal Afar Operation', *Middle East Institute* (10 August 2017): https://www.mei.edu/publications/pmf-forces-reportedly-barred-participating-tal-afar-operation.

Mansour, R. and al Jabar, F. 'The Popular Mobilization Forces and Iraq's Future', *Carnegie Middle East Center* (28 April 2017).

Mansour, R. 'More Than Militias: Iraq's Popular Mobilization Forces Are Here to Stay', *War on the Rocks* (3 April 2018).

Mansour, R. 'The Popularity of the Hashd in Iraq', *Carnegie Middle East Centre* (1 February 2016).

Marsh, N. 'Brothers Came Back with Weapons: The Effects of Arms Proliferation from Libya', *PRISM*, 6:4, 2017, 79-96.

McFate, S. *The Modern Mercenary: Private Armies and What They Mean for World Order* (Oxford: Oxford University Press, 2014).

Moyn, S. *The Last Utopia: Human Rights in History* (Cambridge MA: Harvard University Press, 2012).

Nadimi, F. and Knights, M. 'Iran's Support to Houthi Air Defenses in Yemen', *Policywatch,* 2953 (Washington DC: The Washington Institute for Near East Policy, 4 April 2018).
Nordenman, M. 'The Incredible, Shrinking Modern Military', *The Atlantic* (12 November 2012).
Oakford, S. 'Credibility Gap: United Kingdom civilian harm assessments for the battles of Mosul and Raqqa', *Airwars* (September 2018): https://airwars.org/wp-content/uploads/2018/09/UK-Inquiry-into-Mosul-and-Raqqa-2018.pdf.
'Ruinous Aftermath: Militias Abuses Following Iraq's Recapture of Tikrit', *Human Rights Watch* (20 September 2015).
Sadeghi, P. and Watling, J. 'The Constraints of Infallibility: The Pronouncements of Iran's Supreme Leader', *The RUSI Journal,* DOI: 10.1080/03071847.2020.1767510
Samaha, N., 'Iraq's 'Good Sunni', *Foreign Policy* (16 November 2016).
'Saudi Arabia: Back to Baghdad', *International Crisis Group* (22 May 2018): https://www.crisisgroup.org/middle-east-north-africa/gulf-and-arabian-peninsula/iraq/186-saudi-arabia-back-baghdad.
Singer, P. *Corporate Warriors: The Rise of the Privatized Military Industry* (London: Cornell University Press, 2008).
Smith, P. 'From Karbala to Sayyida Zaynab: Iraqi Fighters in Syria's Shi`a Militias', *CTC Sentinel,* 6:8, 2013, 28-31.
State Sponsors of Terrorism, Bureau of Counterterrorism, US State Department, https://www.state.gov/state-sponsors-of-terrorism/.
Strachan, A. L. 'Factors behind the fall of Mosul to ISIL (Daesh) in 2014', *K4D* (17 January 2014): https://assets.publishing.service.gov.uk/media/59808750e5274a170700002c/K4D_HDR_Factors_behind_the_fall_of_Mosul_in_2014.pdf.
Suleimani, Q., 'Gen. Soleimani congratulates Ayatollah Khamenei and Muslims on ISIS termination', *Khamenei.ir* (21 November 2017): http://english.khamenei.ir/news/5283/Gen-Soleimani-congratulates-Ayatollah-Khamenei-and-Muslims-on.
Terry, F. and McQuinn, B., *The Roots of Restraint in War* (Geneva: ICRC, December 2018).
The Holy Quran, Translated by Abdullah Yusuf Ali (New Delhi: Kitab Bhavan, 2011).
Townsend, C., *The Republic: The Fight for Irish Independence* (London: Allen Lane, 2013).
'UK soldier and two Americans killed in rocket attack in Iraq', *BBC* (12 March 2020).
Watling, J. 'After Effect: The Future Looks Bleak for Iraq Post-Mosul Assault', *Jane's Intelligence Review* (December 2016), 30-33.

5

THE YEMEN CONFLICT AT SEA

Houthi Maritime Operations and Strategic Reconfiguration in the Gulf of Aden

James A. Fargher

Introduction

A military intervention led by Saudi Arabia and the United Arab Emirates (UAE) against Houthi militants in Yemen has precipitated a significant but under-reported redrawing of the geopolitical map surrounding the Red Sea and the Gulf of Aden. In seizing control of the Yemeni state and subsequently voiding its international legitimacy, a violent non-state movement has, in effect, empowered ambitious regional states to expand military reach and political influence beyond the previous confines of recognised borders. By engaging in attacks on international shipping and against civilian targets in Saudi Arabia using suspected Iranian weapons, the Houthis have delegitimised their political project and have invited Saudi Arabia and the UAE to exploit the situation under the mantle of peacekeeping and counterterrorism.

The following chapter will explore how the Houthi rebellion in Yemen has facilitated the expansion of Saudi and Emirati military influence into the globally important shipping route linking the Mediterranean Sea and the Indian Ocean. It seeks to demonstrate that the actions undertaken by an organised group of Violent Non-State Actors (VNSAs) have provided the opportunity and justification for Abu Dhabi and Riyadh to seize islands, purchase naval bases, and form alliances with small local states, jostling their way into an already crowded strategic arena along the Red Sea littoral. The phenomenon itself is not unique; states have used the real or imagined threat posed by VNSAs as a means of obtaining cover for the pursuit of geopolitical interests. Indeed, by challenging existing state-based orders and compromising the sovereignty of territory under their control, VNSAs may open new strategic possibilities previously unavailable due to the restrictions imposed by contemporary norms of international behaviour.

The region surrounding the Red Sea – Gulf of Aden waterway represents a leading example of how VSNAs can reconfigure the local strategic landscape, unintentionally reshuffling the proverbial deck and creating new opportunities for state actors. Following a spate of Houthi advances on Yemen's principal ports, attacks against Emirati, Saudi, and US shipping in the Bab-el-Mandeb Strait, and attempts to mine one of the world's busiest shipping lanes, Saudi Arabia and the UAE have in the name of counter-terrorism established control over a string of harbours, islands, and military bases surrounding the waterway. Whilst the Houthis remain entrenched in northern Yemen and Sanaa despite years of fighting in the Yemeni interior, Saudi Arabia and the UAE have dramatically increased their power projection capabilities into the Red Sea and the Gulf of Aden. Although each new base and the capture of new islands has invariably been justified by each state as a means of countering Houthi threats to maritime commerce, these moves coincide with efforts by other leading regional and maritime states to establish military bases around the Bab-el-Mandeb. These new outposts will augment the Arab states' military posture in the region long after the Yemeni Civil War ends.

This chapter will initially discuss the ways in which state actors can utilise VNSAs in conflict zones to achieve goals around restrictions imposed by state sovereignty. It will then provide an overview of the Houthi rebellion in Yemen and will explore how their toppling of Yemen's internationally recognised government and their attacks on ships off Yemen's coastline have opened new strategic opportunities for Saudi Arabia and the UAE in the Gulf of Aden. The chapter will then turn to the founding of military bases across the Gulf in Djibouti, Eritrea, and Somalia ostensibly to support the intervention in Yemen, noting the wider regional context in which these developments have taken place.

VNSAs and Geopolitics

Since the 11 September 2001 terrorist attacks, states have been repeatedly accused of recasting conflicts against VNSAs to legitimise the pursuit of *realpolitik* goals. As early as 2002, international rights organisation, Human Rights Watch, observed that European governments were scrambling 'to cast their own often brutal internal conflicts as part of the new international anti-terrorist cause'.[1] Russia's campaign against insurgents in Chechnya was a particularly strong example of this, as the Russian government refashioned its ongoing conflict against separatists in the north Caucasus as part of Washington's Global War on Terror – highlighting the presence of foreign terrorist fighters and repeatedly describing the insurgency an attempt to forge an Islamic Caliphate.[2] The Kremlin has adopted a similar line with its more recent military intervention to prevent the collapse of Russia's traditional Middle Eastern ally, the Assad regime in Syria, in which all opposition groups are delegitimised by being tarred as ISIS affiliates.[3]

Similarly, the presence of VNSAs in fragile states is featured prominently in discourse on state failure and the supposed necessity for outside intervention. After the 2001 attacks, a neoconservative argument emerged for increased Western military involvement, potentially even in the form of benign imperialism, to support weak governments challenged by VNSAs.[4] Even when empiric evidence for the relationship between state failure and international terrorism is weak,[5] the image of ungoverned space in an anarchic state harbouring organised terrorist groups remains a potent one. For Western states, 'the discourse of state failure helps to legitimize the spectrum of Western intervention in Africa and other non-Western regions, and in support of Western intervention more broadly'.[6] It is also worth noting that the decision to intervene in any given failed state can also be determined (at least in part) by the apparent strategic value of the

1 Human Rights Watch, 'Europe and Central Asia', *World Report*, 2002.

2 Vladimir Kolossov & Gerard Toal, 'An Empire's Fraying Edge? The North Caucasus Instability in Contemporary Russian Geopolitical Culture', *Eurasian Geography and Economics* 48:2 (2013), 208.

3 Hanna Notte, 'Russia in Chechnya and Syria: Pursuit of Strategic Goals', *Middle East Policy*, 23:1 (Spring 2016), 62.

4 Sebastian Mallaby, 'The Reluctant Imperialist: Terrorism, Failed States, and the Case for American Empire', *Foreign Affairs*, March 2002.

5 Justin George, 'State Failure and Transnational Terrorism: An Empirical Analysis', *Journal of Conflict Resolution* 62:3 (2016): 471-495.

6 Branwen Gruffydd Jones, "Good governance and 'state failure': genealogies of imperial discourse', *Cambridge Review of International Affairs* 26:1 (2013): 49-70.

country in question.[7] In other words, any successful challenge to a weak or otherwise ineffective government of a strategically important state by VNSAs may provide useful cover for outside powers to intervene for their own benefit, with concerns over sovereignty effectively dismissed with the affixation of the 'failed state' label.

The Houthi Insurgency in Yemen

Yemen's position at the southern end of the Arabian Peninsula at the mouth of Bab-el-Mandeb Strait, and its control over several ports and islands lying astride the Suez trade route give it enduring geopolitical importance despite its impoverishment. Whilst the Houthis are perhaps better understood as competitors for control of the Yemeni state rather than opponents to it, the group is routinely labelled a terrorist organisation by Saudi Arabia and its allies.[8] The group's unrecognised control over much of Yemen, affiliation with Iran, and willingness to engage in asymmetric attacks, have allowed outside powers to extend control over Yemeni territory and have legitimised the quiet establishment of permanent garrisons in and around the Gulf of Aden.[9] The Houthis have, in effect, helped Saudi Arabi and the UAE to increase their influence over one of the world's most important trade routes whilst avoiding the controversy faced by China over its conduct in the South China Sea.

The origins of Yemen's ongoing civil war stretch back to 2004 when Houthi militants, drawn primarily from the Zaydi sect based in the northern mountains, began an on-and-off insurgency against the Yemeni government. Formed originally in 1992 in Sanaa as a northern political group operating under the name Shabab al Moumineen, the movement was led by Zaydi cleric Hussein al Houthi before he was killed by Yemeni troops following a siege of his mountain redoubt in September 2004.[10] Although the organisation had lost its core leadership, it mounted an

7 Gabriele Lombardo, 'The responsibility to protect and the lack of intervention in Syria between the protection of human rights and geopolitical strategies', *The International Journal of Human Rights* 19:8 (2015): 1190-1198.

8 Maria-Louise Clausen, 'Justifying military intervention: Yemen as a failed state', *Third World Quarterly* 40:3 (March 2019): 488-502.

9 RT Howard, 'Anything but Strait-Forward: The Saudi Agenda in Yemen', *RUSI Newsbrief* 38:2 (March 2018).

10 Jack Freeman, 'The al Houthi Insurgency in the North of Yemen: An Analysis of the Shabab al Moumieem', *Studies in Conflict & Terrorism* 32:11 (2009), 1009.

intermittent campaign of low-level attacks demanding the overthrow of the Yemeni government. Until 2011, however, the movement remained confined mostly to the Saad'a governate north of Sanaa.

Following the collapse of President Ali Abdullah Saleh's government during the Arab Spring in 2011, the Houthis emerged as a nationally significant force. In 2014, the group broke out of its usual base of operations in Saad'a and captured the capital from the crumbling national government. Declaring the imposition of a revolutionary government for the whole of Yemen, in early 2015 the Houthis subsequently began advancing on Aden, the seat of the internationally recognised Yemeni government under former vice president Abdrabbuh Mansur Hadi.

In response to the impending Houthi capture of Aden, Saudi Arabia announced it would head a military intervention into Yemen with the support of local allies.[11] The campaign's ostensible objective was to eliminate the Houthi movement and to recapture rebel-held territory on behalf of the Yemeni government, in order to crush any Iranian presence in southern Arabia.[12] Accordingly, on 26 March 2015, a coalition of states including Bahrain, Egypt, Jordan, Kuwait, Morocco, Pakistan, Qatar, Sudan and the UAE launched Operation Decisive Storm, in support of Hadi's government. The forces deployed consisted of 30 combat aircraft with intelligence and logistical support from the United States and was accompanied by a full blockade of Yemen's seaports by the Royal Saudi Navy with assistance from the UAE. Somalia, although unable to provide forces to the coalition, did reportedly approve the use of its airspace and port facilities in the northern cities of Bosaso and Berbera for coalition naval and air assets.[13] The coalition formally justified its intervention on the grounds that it had been requested by the internationally recognised Yemeni government and that action would be taken to 'protect the people of Yemen and its legitimate government from a takeover by the Houthis'.[14] Nevertheless, as Roland Popp pointed out in a 2015 policy paper, Operation Decisive Storm fit into a historical pattern of Saudi interference in Yemeni

---

11 Dan Roberts and Kareem Shaheen, 'Saudi Arabia launches Yemen air strikes as alliance builds against Houthi rebels', *The Guardian*, 26 March 2015.

12 Emile Hokayem and David B. Roberts, 'The War in Yemen', *Survival: Global Politics & Strategy* 58:6 (2016), 165.

13 Abdalle Ahmed, 'Somalia lends support to Saudi-led fight against Houthis in Yemen', *The Guardian*, 7 April 2015.

14 Roberts and Shaheen, 'Saudi Arabia launches Yemen air strikes as alliance builds against Houthi rebels'.

domestic politics.[15] It also may have appealed to the Saudi government as a useful opportunity to rally support for the country's new monarch, King Salman.[16]

Despite a sustained month-long aerial campaign, the coalition failed to prevent further Houthi advances and Hadi fled to exile in Saudi Arabia. As early as March 2015, Saudi Arabia and other key partners had indicated a willingness to commit ground troops to Yemen along with warships to prevent continued shipments of arms from Iran to the rebels, which the coalition accused of acting as Tehran's proxies.[17] In April, the UN Security Council issued Resolution 2216 demanding the immediate imposition of a ceasefire and imposing an arms embargo on the country. The resolution not only decided that all member states would take measures to interdict the flow of arms to the Houthis, but also called upon neighbouring states to inspect all shipments bound for Yemeni seaports and airports for contraband.[18] This authorisation for neighbouring states to detain and inspect ships continues to be cited by the Saudi government in support of its ongoing maritime blockade of all Yemeni ports under Houthi control, a move which critics routinely claim has contributed to food scarcity and cholera outbreaks in the country.[19]

Labelling the Houthis as a terrorist group has become an important part of the coalition narrative surrounding the conflict in Yemen, helping to justify military action. In 2016, for example, the military spokesman for the coalition, Brigadier General Ahmed Hassan Al Asiri, described Saudi Arabia's commitment to fighting terrorism throughout the Middle East, lumping the Houthis together with al Qaeda, Hezbollah, and ISIS.[20] A second strand of the coalition's justification for the intervention in Yemen is the assertion that the Houthis operate merely as Iranian proxies, due to the group's acceptance of Iranian weapons.[21] In reality, the true extent of the relationship between the Houthis and Iran is contentious. Alex

15 Roland Popp, 'War in Yemen: Revolution and Saudi Intervention', *CSS Analyses in Security Policy* 175 (June 2015), 4.
16 Stig Stenslie, ''Decisive Storm': Saudi Arabia's attack on the Houthis in Yemen', *NOREF Expert Analysis*, May 2015.
17 Stig Stenslie, 'Decisive Storm'.
18 UN Security Council resolution 2216 (2015), S/RES/2216 (2015) (14 April 2015), available from https://www.un.org/press/en/2015/sc11859.doc.htm.
19 US Library of Congress, Congressional Research Service, *Yemen: Civil War and Regional Intervention*, Jeremy M. Sharp, R43960 (2018), 16.
20 Ahmed Hassan Al Asiri, 'Leading from the Front', *Newsweek Middle East*, 20 April 2016.
21 Khairallah Khairallah, 'Operation Decisive Storm three years on', *The Arab Weekly*, 18 March 2016.

Vatanka at the Middle East Institute casts doubts on the coalition's claims, noting that whilst Iran has supported non-state militants in Yemen since the 1960s, Tehran likely does not view Yemen as anything more than an opportunity to irritate the Saudis, lacking the resources to turn the Houthis into another iteration of Hezbollah.[22] In an analysis of Iran's use of non-state actors in its foreign policy, Thomas Juneau characterised Tehran's relationship with the Houthis as providing modest support in return for extending Iran's influence into Yemen and providing future opportunities to increase pressure on Israel, Saudi Arabia, and the US.[23] On the opposing side, Saudi Arabia and its allies likely viewed the outbreak of war in Yemen as a chance to strike back at Iran by eliminating one of its alleged non-state proxy groups and curtailing its influence in the Arabian Sea.[24] If confirmed, this interpretation would lend further weight to the argument that Saudi Arabia and the UAE have subsequently utilised the presence of the Houthis to boost their long-term regional military presences.

At the time of writing (2020), the Gulf coalition remains embroiled in Yemen despite numerous calls for ceasefires. It is beyond the scope of this chapter to delve into the ground campaign in any great detail, or to provide much commentary on the nature of the Houthi movement itself. Rather, it will focus on how the coalition partners have reacted to Houthi attacks by seizing ports, territory, and the creation of military bases in Yemen and in nearby northeastern Africa. As of late 2019, the Saudi-organised naval blockade of Yemen remains in place, with the port city of Hodeidah still contested.[25] The continued importation of arms by the Houthis into Yemen, their attempts to seize control of islands in the Suez waterway, and their missile attacks on ships off the Yemeni coast are to a substantial degree responsible for turning an intra-state conflict into a regional one. By opening a maritime front against the coalition, the Houthis provided Saudi Arabia and the UAE with a key justification to extend political control over strategic Yemeni islands and to negotiate basing rights agreements with nearby states. These developments will have far-reaching consequences for the regional balance of power.

---

22 Alex Vatanka, 'Iran's Yemen Play: What Tehran Wants – And What it Doesn't', *Foreign Affairs*, 4 March 2015.

23 Thomas Juneau, 'Iran's Policy Towards the Houthis in Yemen: A Limited Return on a Modest Investment', *International Affairs* 92:3 (May 2016), 649.

24 'The percolating proxy war in Yemen', *Strategic Comments*, 23:1 (2017), iv.

25 Ali Mahmood, 'Yemen: Fierce confrontations renewed in Hodiedah', *The National*, 9 March 2019.

The Maritime War

The conflict's naval and coastal operations are particularly noteworthy, because the threatened capture of Yemen's largest harbours by VNSAs and attacks against shipping in the adjacent waterway have resulted in the extension of the coalition partners' permanent military presence in and around the Gulf of Aden. Indeed, as noted by one regional analyst, the Houthi presence in Yemen alone does not adequately explain the extent to which Saudi Arabia and its partners have intervened in the country, as any Iranian proxy threat could have been contained relatively easily.[26] Instead, whilst previously limited by Yemen's status as a sovereign state, Saudi Arabia and the UAE have faced a far lower political cost in gaining control or influence over Yemen's strategic ports. The two Arab powers have effectively blocked any attempt to establish a permanent Iranian naval presence on the coastline, whilst building up their own regional military capacities.

Shortly after the fall of Sanaa, the Houthis began advancing on Yemen's ports and initiated plans to capture the country's islands which dominate the Bab-el-Mandeb waterway linking the Red Sea with the Gulf of Aden. Moreover, the group continued to receive arms shipments from overseas allies, prompting a naval blockade of the country's coastline and the occupation of its principal ports. As early as 8 April 2015, Saudi officials made it clear that the coalition would respond to any attempts by Iranian ships carrying arms for the Houthis to transit through Yemeni territorial waters, later announcing that any vessels attempting to approach the Yemeni coastline would be required to receive prior authorisation from coalition forces.[27] The blockade was later supplemented with amphibious operations to defend or to recapture Yemen's ports from the group. These measures were necessary responses to the Houthi insurgency, which at one point threatened to take Aden, but they have since given Saudi Arabi and the UAE political footholds in Yemen's largest harbours and coastal zones.

One of the coalition's first objectives was to prevent Aden, Yemen's second city and largest port, from falling to the Houthis in April 2015. Aden is one of Yemen's few economic centres, and is a harbour of regional significance, commanding the approaches to the southern entrance of the

26 Howard, 'Anything but Strait-Forward: The Saudi Agenda in Yemen'.

27 Operation Decisive Storm, *Day 21 April 15, 2015: Saudi Ministry of Defense Briefing*, Riyadh: Ministry of Defense, 2015. http://www.operationrenewalofhope.com/day-22-april-15-2015-saudi-ministry-of-defense-briefing/#sthash.oPYRw15c.Pvbs9CUa.dpbs.

Red Sea.[28] Control over Aden would therefore not only have represented a major political coup for the Houthis, cementing their rule over the remnants of the Yemeni state, it would have also given them access to one of the region's only deep-water ports and dockyard. Any violent non-state group in command of Aden would have direct access to the world's trading commons and could pose a significant threat to shipping in the international waterway. The occupation of Aden was conducted initially at the request of the Yemeni government in response to Houthi advances on the city. However, despite moves towards a final political settlement with the group, Aden has remained effectively under coalition control through local actors.[29]

After achieving this immediate objective, Emirati forces were subsequently deployed to seize control of the Hadramout coastline, the long desert shore stretching from Aden to Yemen's eastern border with Oman. A year after repelling the Houthi advance on Aden, the UAE recaptured Mukalla in April 2016 from al-Qaeda in the Arabian Peninsula (AQAP) fighters – another non-state group which capitalised on the breakdown of the state caused by the Houthi's overthrow of Yemen's government.[30] Liberating Mukalla was a key goal given a series of threats issued by AQAP to attack oil tankers attempting to transit nearby – possibly in conjunction with al-Shabaab over in Somalia.[31] As of October 2019, the UAE retains a military and political presence in both Aden and Mukalla – the exact extent of which has not been fully disclosed by the Emirati government.[32] In addition to retaining these garrisons, the UAE turned political control over both ports to the Southern Transitional Council (STC), a coalition of southern non-state separatist groups with which it shares broad cultural similarities and political outlooks.[33] Prior to the collapse of the internationally recognised Yemeni government, this would not have been politically possible. By compromising the sovereignty of Sanaa's rule over

28 Lawrence E. Cline, 'Yemen's Strategic Boxes', *Small Wars Journal* (2009), 9.
29 Ali Mahmood, Yemenis rally in show of support for Saudi-led coalition, *The National*, 6 September 2019.
30 Hind al-Kadimi, 'UAE Arranges its Influence in Yemen's Hadhramount Province', 2 January 2018.
31 Jonathan Saul, 'Al Qaeda in Yemen Poses Growing Threat to Shipping: Naval Force', *Reuters*, 4 May 2016.
32 'UAE pulls some forces from Yemen's Aden as deal nears to end standoff: witnesses', *Reuters*, 9 October 2019; Esraa Ismail & Binsal AbdulKader, 'UAE ship carrying 7,200 tonnes of diesel arrives in Mukalla port', *Emirates News Agency*, 18 October 2019.
33 Adam Baron, 'Foreign and Domestic Influences in the War in Yemen', *Proxy Wars Project Conflict Studies* (2019), 7.

the Hadramout, the Houthis opened the door for the extension of a UAE protectorate over the eastern shoreline.[34]

To the west, in the Bab-el-Mandeb Strait, Houthi successes also prompted the coalition to seize control of strategic territory. In the spring of 2015, militants established control over Perim, the small, barren island which overlooks the southern entrance to the Red Sea. Citing the need to guarantee free passage through the Bab-el-Mandeb Strait, Emirati troops supported by local militia recaptured the island in October 2015, a victory which was hailed by the UAE's foreign minister as 'the beginning of the end of this renegade group and those backing them'.[35] More importantly, however, the Houthis' brief occupation of Perim and threats to attack international shipping effectively licenced an ongoing coalition occupation of the island. The occupying troops remained as a garrison on the island, effectively leaving it under Emirati military administration. After the UAE announced its intention to reduce its military commitments in Yemen, Emirati troops on Perim were replaced with Saudi personnel in July 2019.[36] With tensions escalating between Iran and Saudi Arabia, and given Tehran's known interest in expanding its own naval projection capabilities into the Red Sea,[37] Perim is likely viewed either as a strategic prize by Riyadh, or as too threatening to permit its return and possible recapture by Iranian-aligned forces.

Shortly after Perim's liberation by UAE forces in 2015, Abu Dhabi also began to publicly express interest in extending its influence over the island of Socotra at the eastern mouth of the Gulf of Aden. A month after retaking Perim, Emirati forces were dispatched to Socotra to deliver humanitarian aid following a large cyclone. Despite claims that this move was merely a goodwill gesture towards the Yemeni people, the Emirati presence in Socotra has since remained permanent. Justified as being 'within the efforts of the Arab Coalition to support legitimate government [of Yemen]' against the Houthi rebels, the UAE stationed regular troops at key points on the island, including its only airport, along with allied militia to impose a form of indirect rule.[38] By May 2018, Emirati forces included a significant number

34 Robert Forster, 'The Southern Transitional Council: Implications for Yemen's Peace Process', *Middle East Policy* 24:3 (Fall 2017), 138.

35 Mohammed Mukhashaf, 'Gulf Arabs wrest strategic Yemen island from Iran-allied group', *Reuters*, 5 October 2015.

36 'Saudi Arabia takes over Yemen Red Sea ports after UAE troop withdrawal: Report', *Middle East Eye*, 11 July 2019.

37 'Israel anger at Iran Suez Canal warship move', *BBC News*, 16 February 2011.

38 'UAE says military presence on Yemen's Socotra 'distorted'', *France 24*, 6 May 2018.

of soldiers and armoured vehicles, which were sent over the objections of the Yemeni government in Aden, which described the move as 'an act of aggression'.[39] Perhaps unsurprisingly, by that time the Emiratis were beginning to act less like liberators and more like proconsuls, according to local reports, replacing government flags with UAE ones and local officials with representatives from Abu Dhabi.

In spite of attempts by Saudi Arabia to persuade the UAE to withdraw the bulk of its forces from the island to mitigate Yemeni concerns about its possible annexation, Socotra's airstrip and ports are nevertheless very much still under Emirati control and are expected to remain so.[40] Indeed, reports emerged in early October 2019 of a mutiny staged by UAE-backed fighters who raised the flag of the STC above government buildings after a local official was removed from office on the orders of the Yemeni government in Aden.[41] It would appear that even after abandoning Perim to the Saudis, the UAE retains an interest in extending its military footprint in the middle the Gulf of Aden, and will keep Socotra within its sphere of influence even after any theoretical Houthi threat to the island has abated.

In response to these setbacks and to the ongoing naval blockade, the Houthis launched a series of counterattacks beginning in 2016 against coalition naval forces using retooled ballistic missiles from captured state arsenals and boat-borne Improvised Explosive Devices (IED). As with Houthi moves towards Yemen's ports and islands, these attacks and the likely receipt of technical assistance from Iran have played into a wider narrative espoused by the Gulf coalition which portrays the group as a proxy for Iranian schemes to establish control over both the Strait of Hormuz and the Bab-el-Mandeb.[42]

The first attacks against international shipping by the Houthis occurred in October 2016. On the first day of that month, the UAE reported that its vessel HSV-2 *Swift* had been hit and damaged by a Houthi coastal defence missile fired from near the port of Mokha, prompting condemnation from the UN Security Council.[43] The attack was subsequently claimed by the Houthis, and initial blast damage sustained on the vessel's hull suggested that the ordnance used was a Chinese-manufactured C-802 anti-ship

39 'UAE forces 'occupy' sea and airports on Yemen's Socotra', *Al Jazeera*, 4 May 2018.
40 'UAE withdraws from Yemen's Socotra under Saudi deal', *Middle East Eye*, 18 May 2018.
41 'UAE coup attempt in Yemen's Socotra', *Middle East Monitor*, 5 October 2019.
42 'Saudi halts Red Sea shipment', *The Khaleej Times*, 27 July 2018.
43 US Library of Congress, Congressional Research Service, *Yemen: Recent Attacks Against US Naval Vessels in Red Sea*, Jeremy M. Sharp, IN10599 (2016).

missile, which has been supplied in the past to Iran.[44] The coalition was quick to dub the attack a 'dangerous indication' of the risk posed by the Houthis to international shipping,[45] and in response the US dispatched three warships to monitor the Yemeni coastline.[46]

Shortly afterwards and following a major coalition airstrike in Sanaa which killed several prominent Houthi leaders, between 9 and 12 October militants fired C-802 missiles at US naval vessels in the off the coast, USS *Mason*, *Ponce*, and *Nitze*, prompting retaliatory fire against Houthi-operated radar stations. According to the Pentagon, the Tomahawk strikes were carried out to deter any future 'threat to our ships and commercial traffic, as appropriate, and [the US] will continue to maintain our freedom of navigation in the Red Sea, the Bab el-Mandeb, and elsewhere around the world'.[47] Nevertheless, further attacks against shipping were recorded in January 2017, when a remotely controlled speedboat packed with explosives dubbed a 'drone boat' by some media outlets, targeted a Saudi frigate. In March of that year, a coast guard vessel hit a mine laid by Houthi forces in the Bab-el-Mandeb Strait. Coalition officials reacted to the attacks by reasserting the need to continue naval operations against the Houthis to mitigate against the ongoing risk to 'international navigation and the flow of humanitarian and medical assistance...for Yemeni citizens'.[48]

Following these incidents, the risk posed to international shipping appeared to abate somewhat for the space of a year until April 2018, when the Houthis resumed a spate of drone and missile attacks against civilian shipping in the Red Sea. In April, May and July cargo vessels and tankers, primarily Saudi, were targeted by both IEDs and C-802 missiles, causing the Saudi government to temporarily halt oil shipments through the Bab-el-Mandeb. The attacks appear to have been orchestrated as part of a previous threat issued by the Houthis to close the Bab-el-Mandeb entirely should the coalition fail to lift the blockade on Hodeidah, Yemen's second largest port located on its western coast.[49] Indeed, following the initial attacks in April and May 2018, Emirati foreign minister Anwar Gargash published an editorial in

44 Sam LaGrone, 'USS Mason Fired 3 Missiles to Defend from Yemen Cruise Missiles Attack', *US Naval Institute News*, 11 October 2016.

45 'Houthis target UAE ship carrying aid', *Al-Arabiya*, 1 October 2016.

46 Sharp, *Yemen: Recent Attacks Against US Naval Vessels in Red Sea.*

47 US Department of Defense, 'Statement by Pentagon Press Secretary Peter Cook on US Military Strikes Against Radar Sites in Yemen', *News Release*, 12 October 2016.

48 'Two killed after Houthis attack Saudi warship', *Al-Arabiya*, 31 January 2017.

49 Aziz El Yaakoubi, 'Yemen's Houthis threaten to block Red Sea shipping lane', *Reuters*, 9 January 2018.

The Telegraph explaining that the coalition would continue to contest Houthi control over Hodeidah. Gargash described the Houthis' continued presence in Hodeidah as 'a clear and present danger to the strategic security of international shipping lanes from the Bab el Mandeb Straights [sic] through the Red Sea to the Suez Canal', and threats to close the waterway as 'a criminal threat…to gain leverage over the international community'.[50] The statement made it clear that the coalition would attempt to wrest control of the port as a means of protecting the international shipping lane and that naval operations off the Yemeni coast would continue as part of the ongoing operation.

Nevertheless, failing to quickly seize Hodeidah from entrenched Houthi fighters, the coalition suffered further attempted attacks by Houthi IEDs in August 2018. On 29 September 2018, the Royal Saudi Navy reported intercepting two Houthi speedboats packed with explosives bound for the Saudi port of Jazan. Incidents of piracy have also continued to occur in the waterway, including suspicious skiffs approaching internationally flagged vessels in the Gulf of Aden on four separate occasions in November and December 2018, and February 2019. Although it is not clear whether these boats were operated by Houthi militants, or were piloted by criminal groups, they contribute to a wider depiction of regional maritime insecurity cited by the coalition as a need for extensive military operations off the Yemeni coast. Similarly, the routine discovery of sea mines and improvised torpedoes in the Red Sea are attributed to Houthi asymmetrical counter-attacks against the naval blockade. In October 2018, for example, following the release of footage by the group displaying manufactured mines, regional media outlets described the existence of these mines as 'a flagrant to the Red Sea shipping routes, and further proof of the Iranian involvement'.[51] Observers and Saudi officials have also since linked the attacks with threats issued routinely by Tehran against Washington to close the Strait of Hormuz, suggesting Iran may be pursuing a two-pronged strategy to exert control over both strategic chokepoints.[52] Iran has certainly signalled its interest in obtaining the strategic capability to interdict oil shipments through the Bab-el-Mandeb, although the extent of the Houthi-Iranian relationship remains something of an open question.[53]

---

50 Anwar Gargash, 'To take on the Houthi rebels in Yemen, we must liberate Hodeidah', *The Telegraph*, 14 June 2018.
51 Ali Rabih, 'Yemen: Houthis Openly Plant Naval Mines', *Asharq Al-Awsat*, 19 October 2018.
52 Keith Johnson, 'Iran's Yemeni Proxies Put Oil Shipments in Crosshairs', *Foreign Policy*, 26 July 2019; Adel Aljubeir, Twitter post, 11 September 2018, 12:46pm.
53 Saeed Kamali Dehghan and Patrick Wintour, 'Senior Iran military chief tells Trump: 'We're Ready to Stand Up to You'', *The Guardian*, 26 July 2019.

Attacks against shipping in the Red Sea have also coincided with a spate of ballistic missile strikes launched by militants into Saudi Arabia, along with attempted drone attacks on Dubai International Airport and a claimed attack on Saudi Aramco oil facilities in September 2019. Despite ongoing mediation efforts, the coalition has warned that the risk of further attacks in the Bab-el-Mandeb is likely.[54] Both Saudi Arabia and the UAE accused the 'Iran-backed Houthis' of violating the Stockholm Agreement and Saudi forces continue to contest Hodeidah, claiming that liberating the port from Houthi control is essential for the safety of international shipping.[55] Although the extent of the relationship between Iran and the Houthis remains uncertain,[56] by conducting these attacks on nearby vessels, the Houthis have enabled Saudi Arabia and the UAE to extend control over strategic sections of sovereign Yemeni territory. Moreover, the alleged militant threat to the international shipping lane has also proven to be a key factor in both states' pursuit of military bases across the gulf in northeastern Africa.

Imperial Expansion: A New Scramble

Houthi actions in Yemen and the Suez waterway proved to be a catalyst for a significant recalibration of the region's strategic picture as both Abu Dhabi and Riyadh have focused on the African shores of the Red Sea and the Gulf of Aden for military bases and political alliances. Notably, whilst the conflict continues to drag on in the Yemeni interior, coalition moves to secure the coastline and islands occurred at an early stage of the intervention. Combined with a competition for bases in the Gulf of Aden by outside powers, these developments echo a similar struggle for supremacy waged by the European maritime empires in the late nineteenth-century. In this context, the need to prevent attacks from non-state actors labelled as extremists presents an uncontroversial motive for acquiring these bases. The immediate response to Houthi attacks in Yemen may well have required the need for temporary interventions along the Yemeni coast. But

54 Robin Mills, 'Strait of Hormuz is not the most dangerous risk for blockade', *The National*, 31 July 2018.
55 Patrick Wintour, 'Saudi-led forces launch airstrikes on Yemeni city of Hodeidah', *The Guardian*, 20 September 2019.
56 UN Security Council, 'Final Report of the Panel of Experts on Yemen', S/2019/83, 25 January 2019.

beginning with Socotra, the coalition allies began referring to the Houthi threat as justification for an ever-growing expansion of their power in and around the Gulf of Aden. Africa, Saudi Arabia and the UAE continue to cite the Houthis as a reason for establishing a string of military bases, even whilst clearly seeking an exit from Yemen.[57] By engaging in attacks on coalition maritime forces, the Houthis now act as spectres needed to disguise the coalition's pursuit of old-fashioned *realpolitik*, which threatens to complicate an already tangled regional outlook.

Saudi Arabia's efforts to expand its military presence into the region have so far concentrated on developing a base in Djibouti at the southern mouth of the Bab-el-Mandeb. Formal discussions regarding the base were first opened less than a year after the initial intervention in Yemen and two months after the country severed ties with Iran – likely at Riyadh's request or in the hopes of fostering improved relations with the kingdom. In March 2016, Djibouti's ambassador to Riyadh, Ziauddin Bamakhrama, confirmed in a media interview that the two countries were negotiating the opening of a base to host Saudi air, naval and ground units. Significantly, Bamakhrama explicitly referenced Djibouti's role in blocking 'Iran's attempts to expand the influence of the Houthi militias' as a factor in aligning with Saudi Arabia.[58] President Ismail Omar Guelleh himself linked the proposed base with attacks on shipping in the Bab-el-Mandeb Strait, describing the proposed facility as a resource to 'support for our navy and coast guard forces…in order to carry out our local and regional role which has been entrusted to us to protect our territorial waters first and then international waters'.[59]

Although presented as a means of countering the Houthi maritime threat, in subsequent media interviews, Djiboutian officials revealed that the opening of the Saudi base will take place as part of a much broader scheme to align the two country's strategic interests. Djibouti defence minister Ali Bahdon in November 2017, for example, was quoted saying that 'it is a comprehensive strategic cooperation between the two countries in various military aspects, not in a specific sector and role'.[60] Media outlets close to Djibouti government spokesmen have noted that the base, when it

57 Mohammed Hatem, 'Saudi Arabia Agrees to Limited Cease-Fire in Yemen War', *Bloomberg*, 27 September 2019.
58 Fahd al-Zayabi, 'Djibouti: We Welcome the Establishment of a Saudi Military Base on Our Territory', *Asharq al-Awsat*, 8 March 2016.
59 Fahd al-Zayabi, 'Djibouti'.
60 Ali Bahdon, quoted in Saeed al-Abyadh, 'Djibouti's Defense Minister to Ashraq Al-Awsat: We Welcome Saudi Military Base', *Ashraq Al-Awsat*, 28 November 2017.

opens, will give Saudi Arabia greater capabilities to control the suspected flow of weapons from Iran to the Houthis.[61] Indeed, Iranian state media has portrayed the base as part of a wider scheme for Riyadh to acquire regional influence, describing it variously as an 'intelligence outpost, through which it can monitor Iran's purported movements in the region' and as 'a base from which it can thoroughly dominate the strategic Bab-el-Mandeb Strait in a bid to exert more pressure on Yemen's Houthi Anasrullah movement'.[62] Whilst the complement to be deployed to the future base has not yet been announced, it will presumably host the air and maritime facilities capable of extending Saudi influence into the Gulf of Tadjourra.

Regional scholar Alex de Waal has argued that Saudi Arabia's moves into Djibouti have come as part of a wider strategy to expand Saudi influence into a strategic hinterland, in part to block the ambitions of an emerging Turkish-Qatari-Iranian axis in the region.[63] As he notes, the Gulf monarchies were indeed involved in attempts to restrict and counter Iranian influence in the Horn of Africa in the early 2000s prior to the outbreak of the conflict in Yemen. However, the timing of events clearly suggests that it was the Saudi decision to intervene militarily in Yemen, so often defended as imperative to secure the Red Sea from Iranian proxies, which spurred Riyadh into opening its first overseas base at the mouth of the Bab-el-Mandeb.

For its part, the UAE has proven to be much more ambitious in the territorial scope of its mission to establish new ports and military bases than its Saudi partner. In Africa, the UAE has opened a number of naval bases in order to support its operations against the Houthis. In the summer of 2015, it signed an agreement with Eritrea to develop Assab, a formerly small and impoverished town originally founded as Italy's first colonial settlement. The UAE has subsequently made significant investments in the port, developing it as a support base for its operations against the Houthis and as a site to train allied special forces and militia, as well as to possibly host off-shore detention facilities. An expert report submitted to the UN Security Council in November 2018 sketched out the extent of the Emirates' development of Assab as a forward operating base. According to the

61 Sahwqi al-Rayyes, 'Djibouti on the Rise as Hub for Foreign Military Bases in Africa', *Asharq al-Awsat*, 10 September 2018.
62 'Djibouti hails Saudi military presence in African country as way to boost bilateral ties', *PressTV*, 28 November 2017.
63 Alex de Waal, 'Pax Africana or Middle East Security Alliance in the Horn of Africa and the Red Sea?', *World Peace Foundation, Occasional Paper* 71 (January 2019).

report, by late 2017 docks capable of handling blue-water warships had been constructed, as well as an airbase suitable for long-range unmanned aerial vehicles, possibly the Chinese Wing Loong II.[64] UN inspectors also indicated that Abu Dhabi has shipped large quantities of weapons to the base, potentially to be distributed to local affiliates, although these claims were denied by the Emirati government.[65] It is worth noting that the base at Assab, which was leased to the Emirates for a period of 30 years, has since become the country's key site for expanding its military footprint in the Red Sea, despite being initially described as a facility to support Emirati efforts to defend Aden from advancing Houthi forces.

In February 2017, Abu Dhabi also negotiated a 30-year lease on the port of Berbera, the largest harbour along the Somali coastline on the Gulf of Aden, which was followed by another lease on the small port of Boosaaso in October 2017. The Somali federal government based in Mogadishu initially agreed to permit the coalition to use Somalia airspace following the outbreak of hostilities with the Houthis in April 2015.[66] However, Abu Dhabi opened negotiations for control over Berbera and Boosaaso in May 2016 with the unrecognised breakaway governments of Somaliland and Puntland, which lack any international recognition. The move infuriated the Somali government but underlined the lengths to which the UAE has used the opportunity provided by the war in Yemen to maximise its own position without attracting the opprobrium of the Great Powers.

Although ostensibly carried out by DP World, a commercial shipping firm controlled by the Emirati government, the implications of Emirati ownership of both ports are clear.[67] The UAE has already begun operations to establish a naval facility next to the port of Berbera, giving its ships a permanent base on the southern end of the Bab-el-Mandeb in addition to its position north of the strait at Assab. The base is expected to be completed by June 2019, and will feature two airstrips, a dockyard and a coastal surveillance system, a significant expansion of Emirati power in the gulf. Somaliland officials have admitted forthrightly that this base will serve a far

64 UN Security Council, 'Letter dated 7 November 2018 from the Chair of the Security Council Committee pursuant to resolutions 751 (1992) and 1907 (2009) concerning Somalia and Eritrea addressed to the President of the Security Council', S/2018/1003, 9 November 2018.

65 UN Security Council.

66 Abdalle Ahmed, 'Somalia lends support to Saudi-led fight against Houthis in Yemen', *The Guardian*, 7 April 2015.

67 David Styan, 'The politics of the ports in the Horn: War, peace and Red Sea rivalries', *African Arguments*, 18 July 2018.

more long-lasting purpose than merely supporting military operations in Yemen.[68]

Conclusion

The Houthi rebellion in Yemen provides a useful example of how conflicts with VNSAs complicate the wider political environment and can open opportunities for involved state actors beyond the strictures of Westphalian order. By seizing control of state institutions and dominating large stretches of territory, the Houthis compromised Yemen's sovereignty and permitted the expansion of the Saudi and Emirati military footprint beyond what may have been politically feasible prior to the conflict. This is perhaps most obvious with the annexation of Yemeni islands and ports, but also with the less-publicised founding of coalition military bases in Djibouti, Eritrea, and Somalia which has been carried out without sparking a significant degree of international controversy. Capitalising on the overthrow of the Yemeni government by the Houthis, Saudi Arabia and the UAE are now left with permanent garrisons in Yemen, or in the case of Perim and Soctora, with virtual protectorates. Although a roadmap to a final political settlement in Yemen still appears far off, Saudi and Emirati behaviour towards these strategic points suggests that they will continue to host coalition military forces even after the conclusion of hostilities.

In Africa, this trend is even more explicit. Citing the need to support operations against the Houthis, the coalition has founded bases in Djibouti, Eritrea, and Somaliland which appear to support long-term strategic ends rather than providing immediate tactical solutions for fighting in Yemen. Regardless of the outcome of the war in Yemen, these bases will give the Gulf monarchies the ability to project naval and air power over the strategic shipping route and the Horn of Africa for decades – which, this chapter suggest, is the true intention behind their founding. Indeed, the extent of the leases negotiated on the bases and the considerable level of development being poured into the sites certainly support this conclusion. Although the Saudi and Emirati governments have been reluctant to publicly discuss the future of the bases in any great detail, the information available suggest that will become key aspects of the region's future security architecture in

68 Nizar Manek, 'UAE Military Base in Breakaway Somaliland to Open in June', *Bloomberg*, 6 November 2018.

the wake of waning US interest in the Middle East. Looking ahead, this may well carry significant implications for Western and Middle Eastern policymakers alike as the regional priorities for both coalition partners as well as the US and European allies slowly begin to diverge.

References

Ahmed, A. 'Somalia lends support to Saudi-led fight against Houthis in Yemen'. *The Guardian*, 7 April 2015.

Aljubeir, A. *Twitter post*, 11 September 2018.

Al-Asiri, A.H. 'Leading from the Front'. *Newsweek Middle East*, 20 April 2016.

Al-Kadimi, H. 'UAE Arranges its Influence in Yemen's Hadhramount Province'. 2 January 2018.

Al-Karimi, K. 'UAE offers a helping hand to the island of Socotra'. *The National*, 16 April 2017.

Al-Rayyes, S. 'Djibouti on the Rise as Hub for Foreign Military Bases in Africa'. *Asharq al-Awsat*, 10 September 2018.

Al-Zayabi, F. 'Djibouti: We Welcome the Establishment of a Saudi Military Base on Our Territory'. *Asharq al-Awsat*, 8 March 2016.

'Arrival of UAE-backed forces stokes tensions on Yemen's Socotra'. *Middle East Eye*, 4 July 2019.

Baron, A. 'Foreign and Domestic Influences in the War on Yemen'. *Proxy Wars Project Conflict Series* (2019): 1-9.

Binnie, J. 'UAE stops work on Bab al-Mandab island base'. *Janes Defence Weekly*, 28 March 2018.

Clausen, M. 'Justifying military intervention: Yemen as a failed state'. *Third World Quarterly* 40:3 (March 2019): 488-502.

Cline, L.E. 'Yemen's Strategic Boxes'. *Small Wars Journal* (2009): 1-12.

De Waal, A. 'Pax Africana or Middle East Security Alliance in the Horn of Africa and the Red Sea?' *World Peace Foundation*. Occasional Paper 71 (January 2019).

Dehghan, S. E. and P. Wintour. 'Senior Iran military chief tells Trump: "We're Ready to Stand Up to You"'. *The Guardian*, 26 July 2019.

'Djibouti hails Saudi military presence in African country as way to boost bilateral ties'. *PressTV*, 28 November 2017.

Forster, R. 'The Southern Transitional Council: Implications for Yemen's Peace Process'. *Middle East Policy* 24:3 (Fall 2017): 133-144.

Frantzman, S.J. 'Socotra: How a Strategic Island Became Part of a Gulf Power Struggle'. *Jerusalem Post*, 6 May 2018.

Freeman, J. 'The al Houthi Insurgency in the North of Yemen: An Analysis of the Shabab al Moumieem'. *Studies in Conflict & Terrorism* 32:11 (2009): 1008-1019.

Gargash, A. 'To take on the Houthi rebels in Yemen, we must liberate Hodeidah'. *The Telegraph*, 14 June 2018.

George, J. 'State Failure and Transnational Terrorism: An Empirical Analysis'. *Journal of Conflict Resolution* 62:3 (2016): 471-495.

Hatem, M. 'Saudi Arabia Agrees to Limited Cease-Fire in Yemen War'. *Bloomberg*, 27 September 2019.

Hokayem, E. and D. B. Roberts, 'The War in Yemen'. *Survival: Global Politics & Strategy* 58:6 (2016): 157-186.

'Houthi Rebels Threaten to Bomb Somaliland's Berbera Port Run by UAE'. *Shabelle*, 25 December 2017.

'Houthis target UAE ship carrying aid'. *Al-Arabiya*, 1 October 2016.

Howard, R.T. 'Anything but Strait-Forward: The Saudi Agenda in Yemen'. *RUSI Newsbrief* 38:2 (March 2018).

Human Rights Watch. 'Europe and Central Asia'. *World Report*, 2002.

Ismail, E. and B. A. Kader. 'UAE ship carrying 7,200 tonnes of diesel arrives in Mukalla port'. *Emirates News Agency*, 18 October 2019.

Johnson, K. 'Iran's Yemeni Proxies Put Oil Shipments in Crosshairs'. *Foreign Policy*, 26 July 2019.

Juneau, T. 'Iran's Policy Towards the Houthis in Yemen: A Limited Return on a Modest Investment'. *International Affairs* 92:3 (May 2016): 647-663.

Khairallah, K. 'Operation Decisive Storm three years on'. *The Arab Weekly*, 18 March 2016.

Kolossov, V. & G. Toal. 'An Empire's Fraying Edge? The North Caucasus Instability in Contemporary Russian Geopolitical Culture'. *Eurasian Geography and Economics* 48:2 (2013): 202-225.

LaGrone, S. 'USS Mason Fired 3 Missiles to Defend from Yemen Cruise Missiles Attack'. *US Naval Institute News*, 11 October 2016.

Lombardo, G. 'The responsibility to protest and the lack of intervention in Syria between the protection of human rights and geopolitical strategies'. *The International Journal of Human Rights* 19:8 (2015): 1190-1198.

Mahmood, A. 'Yemen: Fierce confrontations renewed in Hodiedah'. *The National*, 9 March 2019.

Mahmood, A. 'Yemenis rally in show of support for Saudi-led coalition'. *The National*, 6 September 2019.

Mallaby, S. 'The Reluctant Imperialist: Terrorism, Failed States, and the Case for American Empire'. *Foreign Affairs*, March 2002.

Manek, N. 'UAE Military Base in Breakaway Somaliland to Open in June'. *Bloomberg*, 6 November 2018.

Mills, R. 'Strait of Hormuz is not the most dangerous risk for blockade'. *The National*, 31 July 2018.

Mukhashaf, M. 'Gulf Arabs wrest strategic Yemen island from Iran-allied group'. *Reuters*, 5 October 2015.

National Archives. Kew. Foreign Office Confidential Records: Correspondence Respecting the Red Sea and Somali Coast. Letter from Sir J. Pauncefort to Mr. Godley, 20 July 1885.

Notte, H. 'Russia in Chechnya and Syria: Pursuit of Strategic Goals'. *Middle East Policy*, 23:1 (Spring 2016): 59-74.

Operation Decisive Storm, *Day 14 April 2015*: Saudi Ministry of Defense Briefing, Riyadh: Ministry of Defense, 2015. http://www.operationrenewalofhope.com/day-15-april-8-saudi-ministry-of-defense-briefing/#sthash.m0EMMzat.dpbs.

Operation Decisive Storm, *Day 21 April 15, 2015*: Saudi Ministry of Defense Briefing. Riyadh: Ministry of Defense, 2015. http://www.operationrenewalofhope.com/day-22-april-15-2015-saudi-ministry-of-defense-briefing/#sthash.oPYRw15c.Pvbs9CUa.dpbs.

'The percolating proxy war in Yemen'. *Strategic Comments*, 23:1 (2017): iv-vi.

Popp, R. 'War in Yemen: Revolution and Saudi Intervention'. *CSS Analyses in Security Policy*, 175 (June 2015): 1-4.

Rabih, A. 'Yemen: Houthis Openly Plant Naval Mines'. *Asharq Al-Awsat*, 19 October 2018.

Ramm, A. 'Great Britain and the Planting of Italian Power in the Red Sea, 1868 – 1885'. *The English Historical Review* 59:234 (May 1944): 211-236.

Roberts, D. and K. Shaheen. 'Saudi Arabia launches Yemen air strikes as alliance builds against Houthi rebels'. *The Guardian*, 26 March 2015.

'Saudi Arabia takes over Yemen Red Sea ports after UAE troop withdrawal: Report'. *Middle East Eye*, 11 July 2019.

'Saudi halts Red Sea shipment'. *The Khaleej Times*, 27 July 2018.

Saul, J. 'Al Qaeda in Yemen Poses Growing Threat to Shipping: Naval Force'. *Reuters*, 4 May 2016.

Stenslie, S. "Decisive Storm': Saudi Arabia's attack on the Houthis in Yemen'. *NOREF Expert Analysis*, May 2015.

Styan, D. 'The politics of the ports in the Horn: War, peace and Red Sea rivalries'. *African Arguments*, 18 July 2018.

Telci, I. and T. Horoz. 'Military Bases in the Foreign Policy of the United Arab Emirates'. *Insight Turkey* 20:2 (2018): 143 – 165.

'Two killed after Houthis attack Saudi warship'. *Al-Arabiya*, 31 January 2017.

'UAE forces 'occupy' sea and airports on Yemen's Socotra'. *Al Jazeera*, 4 May 2018.

'UAE pulls some forces from Yemen's Aden as deal nears to end standoff: witnesses'. *Reuters*, 9 October 2019.

'UAE says military presence on Yemen's Socotra 'distorted'. *France24*, 6 May 2019.

'UAE withdraws from Yemen's Socotra under Saudi deal'. *Middle East Eye*, 18 May 2018.

UN Security Council resolution 2216 (2015). S/RES/2216 (2015) (14 April 2015). https://www.un.org/press/en/2015/sc11859.doc.htm.

UN Security Council. 'Final Report of the Panel of Experts on Yemen'. S/2019/83, 25 January 2019.

UN Security Council. 'Letter dated 7 November 2018 from the Chair of the Security Council Committee pursuant to resolutions 751 (1992) and 1907 (2009) concerning Somalia and Eritrea addressed to the President of the Security Council'. S/2018/1003, 9 November 2018.

UN Security Council. 'Letter from the Permanent Representatives of Saudi Arabia, the United Arab Emirates and Yemen to the United Nations addressed to the President of the Security Council'. S/2019/99, 1 February 2019.

US Department of Defense. 'Statement by Pentagon Press Secretary Peter Cook on US Military Strikes Against Radar Sites in Yemen'. *News Release*, 12 October 2016.

US Library of Congress. Congressional Research Service. *Yemen: Civil War and Regional Intervention*, Jeremy M. Sharp. R43960 (2018).

US Library of Congress. Congressional Research Service. *Yemen: Recent Attacks Against US Naval Vessels in Red Sea*. Jeremy M. Sharp, IN10599 (2016).

Vatanka, A. 'Iran's Yemen Play: What Iran Wants – And What it Doesn't'. *Foreign Affairs*, 4 March 2015.

Wintour, P. 'Saudi-led forces launch airstrikes on Yemeni city of Hodeidah'. *The Guardian*, 20 September 2019.

Yaakoubi, A. 'Yemen's Houthis threaten to block Red Sea shipping lane'. *Reuters*, 9 January 2018.

6

WARFARE BY PROXY

Russia's Use of Violent Non-State Actors in Syria

Marina Miron

Introduction

The conflict in Syria, a product of the Arab Spring, commenced in 2011. It began as the struggle of Bashir Al Assad's government forces to maintain control in the face of a variety of internal opposition movements. It quickly came to also involve a number of external actors. It became, indeed, the main focal point for what was seen by certain NATO states, and especially by the United States (US), as the global struggle against the Islamic State (IS).[1] The situation was complex. What gave this conflict its particularly Byzantine character was not just the plethora of clashing interests of both domestic and international actors but also the widespread use of Violent Non-State Actors (VNSAs).

VNSAs are understood here as entities such as private militias, warlords, tribes and private military contractors. Some can be of a hybrid character, combining several of the above.[2] Their decision-making and behaviour are mainly influenced by the struggle for territorial and political control within the environment of 'weak' states with fragile political,

1 Samuel Charap, 'Russia, Syria and the Doctrine of Intervention', *Survival*, 55:1 (2013), 37.
2 Troy S. Thomas and William D. Casebeer, *Violent Systems: Defeating Terrorist, Insurgents, and Other Non-State Adversaries*. (Colorado Springs: USAF Institute for National Security Studies, 2004) 4.

socioeconomic and security frameworks.[3] VNSAs can also, of course, often present a viable proxy for third-party state actors intervening in a particular conflict and who wish to further their own political and strategic interests without the risk of a full-scale commitment of their own.

The Syrian situation can be perceived as a classic example of what Andrew Mumford sees as 'proxy warfare'; that is, 'the indirect engagement in a conflict by third parties seeking to influence its strategic outcome'.[4] In this Syrian case there has existed a multi-layered proxy war wherein a host of VNSAs[5] has been – in a broad sense – sponsored, trained and equipped by another host of external players. Included here are the US, Russia, Turkey, Iran and certain Arab states.

Notwithstanding its complexity, after eight years of intense struggle, the conflict in Syria is now reducing in scale.[6] And for some of the third-party states, Syria is no longer at the forefront of their geopolitical agenda and their 'strategic outcome' has been achieved.[7] In October 2019, for instance, the Trump administration made the decision to withdraw from Syria altogether, stating that 'the United States has defeated the territorial caliphate [of the IS]'.[8] The other major external actor in Syria – Russia – is, though, still there. Moscow, indeed, seems to have emerged as the 'big winner'.[9]

The events in the Syrian theatre raise some important questions for policy makers in NATO member states which might be engaged in future, similar conflicts. Russia, overall, used its proxies adroitly. This chapter

3 See, for instance, Robert H. Jackson, *Quasi-Sates: Sovereignty, International Relations, and the Third World.* (Cambridge: Cambridge University Press, 1990); Sunil Dasgupta, *Paramilitary Forces and Security Reorganization*, GSC Quarterly. Program on Global Security and Cooperation, Social Science Research Council (2004), available at: http://www.ssrc.org; Stuart E., Eizenstat, John Edward Porter, and Jeremy M. Weinstein, 'Rebuilding Weak States', *Foreign Affairs*, 84:1 (2005), 134–146; Alexus G. Grynkewich, 'Welfare as Warfare: How Violent Non-State Groups Use Social Services to Attack the State', *Studies in Conflict & Terrorism*, 31:4 (2008), 350–370; and Robert J. Bunker, 'Defeating Violent Nonstate Actors', *Parameters*, 43:4 (2013), 57-65.

4 Andrew Mumford, *Proxy Warfare*. (Cambridge, UK: Polity Press, 2013), 12.

5 For an approximate list of VNSAs in Syria, see Cody Roche, 'Syrian Opposition Factions in the Syrian Civil War', *Bellingcat*, August 13, 2016, https://www.bellingcat.com/news/mena/2016/08/13/syrian-opposition-factions-in-the-syrian-civil-war/.

6 'Voyna v Sirii Ne Zakonchena, Zayavil Lavrov [The War in Syria Has Not Ended, Said Lavrov]', *RIA*, April 3, 2019, https://ria.ru/20190403/1552359037.html.

7 Mumford, *Proxy Warfare*, 12.

8 'Trump Orders US Troops Out of Northern Syria as Turkish Assault Continues', *The Guardian*, October 13, 2019, https://www.theguardian.com/world/2019/oct/13/trump-us-troops-northern-syria-turkish-assault-kurds.

9 The Associated Press, 'Why Russia is the Big Winner from Trump's Syria Pullback and Ukraine Turmoil', *Haaretz*, October 13, 2019, https://www.haaretz.com/us-news/why-russia-is-the-big-winner-from-trump-s-syria-pullback-and-ukraine-turmoil-1.7970597.

shows how Moscow has successfully managed to employ VNSAs in a very complex environment and to achieve what may be seen as its objectives – to have gained its 'strategic outcomes'. These include keeping its ally in power by helping suppress the IS and the opposition and establishing itself as one of the key players and power brokers in the region. The chapter proceeds to discuss two specific VNSAs used by Russia: the Syrian-Kurdish *Yekîneyên Parastina Gel* (People's Protection Units or YPG), and the private military company (PMC), the Wagner Group. The following sections look at Russia's objectives in Syria and discuss the employment of each VNSA not only in military but also in strategic terms in order to understand the underlying impetus for their utility and possible shortcomings their use might have had.

Russia's Objectives in Syria

Before considering how and why Russia used VNSAs during its Syrian campaign, it is necessary first to understand Moscow's overall role in the conflict and to briefly discuss its objectives. This allows for a situating of the role of the VNSAs it employed within the overarching strategy pursued by the Kremlin.

Russia's military engagement in Syria – commencing in September 2015 – was the result of several drivers. The geostrategic location of Syria, coupled with its long-standing importance as an ally for the former Soviet Union, provides some clues as to why Russia intervened. During the Cold War, Syria and Libya had been long-standing allies of the Soviet Union and had provided bases enabling Moscow to project power in the Middle East and North African (MENA) region.[10] After the fall of the Soviet Union, however, Russia had reached its economic nadir. It was no longer a superpower. Allies such as Syria and Libya had to be abandoned. When Vladimir Putin succeeded Boris Yeltsin as leader in 1999, he sought to re-establish Russia as a global player, including restoring close ties with former allies.[11] This was an essential move propagated by a key adviser

10 John M. Mackintosh, *Strategy and Tactics of Soviet Foreign Policy*. (New York: Oxford University Press, 1963), 117-128; 242pp. 'Punkt Material'no-Tekhnicheskogo Obezpecheniya VMF Rossii v Tartuse. Dos'ye [Logistical Point of the Russian Navy in Tartus. Dossier]', *TASS*, December 13, 2017, https://tass.ru/info/4808523.

11 Vladimir Putin, 'Poslaniye Federal'nomy Sobraniyu Rossiiskoi Federacii [Message to the Federal Assembly of the Russian Federation]', *Kremlin*, April 25, 2005, http://kremlin.ru/events/president/transcripts/22931.

to Putin, Alexandr Dugin's (known as 'Putin's Rasputin').[12] Dugin's interpretation of geopolitics suggested that gaining allies would help Russia not only to reassert its global position, but also act to protect Russia's national security interests.[13] And certainly, following the NATO-facilitated regime change in 2011 in Libya (when Muammar Ghaddafi, an old friend of Moscow, was removed), there emerged fears that Russian influence in the whole MENA region could fall victim to what was perceived as Western adventurism.[14] Hence, saving Syria, that is, its leader Assad, from a similar fate became a central aim of Russian geopolitical thinking and one of its prime strategic goals post-2011.[15] Another strategic goal had also emerged by 2015: the defeat of the IS. Left unchecked, the Kremlin feared IS-precipitated conflict beyond Syria's borders and eventually into Russia's southern tier of Muslim-majority republics including Chechnya, Dagestan and Ingushetia.[16] This further reinforced Moscow's resolve to intervene.

Russia's campaign in Syria had thus an initial two-fold purpose. In the long term, Russia sought to save its ally and deny the West yet another regime change in one of Moscow's traditional spheres of interest. In the short-term, the Kremlin feared that the jihadism present in Syria could potentially reach Russia, threatening its internal stability. In short, Russia had distinct reasons to stay committed to Syria in ways that Western states were not. This resolve certainly came to influence both the choice of Russian VNSAs in Syria and the ways in which it utilised them.

---

12 Amos Barshad, 'Putin's Rasputin', *Longread,* July 11, 2019, https://longreads.com/2019/07/11/putins-rasputin/.

13 Alexandr Dugin, *Osnovy Geopolitiki: Geopoliticheskoe Budushhee Rossii* [The Foundations of Geopolitics: Russia's Geopolitical Future]. (Moscow: Arktogeja, 1997), 91-93.

14 Artem Krechetnikov, 'Medvedev y Putin: Govorim Liviya, Podrozumevayem Kreml' [Medvedev and Putin: Saying Libya, Thinking Kremlin]', *BBC,* March 22, 2011, https://www.bbc.com/russian/russia/2011/03/110322_medvedev_putin_libya; Rod Thornton, 'The Russian Military's New "Main Emphasis"', *The RUSI Journal,* 162:4 (2017), 20-21.

15 'Putin Nazval Osnovnuyu Zadachu Rossiyskih Voennyh v Sirii [Putin Named the Main Aim for the Russian Military in Syria]', *Interfax,* October 11, 2015, https://www.interfax.ru/russia/472593.

16 Sergei Val'chenko, 'Eks-Glavkom VKS Podvel Itogi Operacii v Sirii za Tri Goda [Former Commander-in-Chief of the VKS Summed Up The Results of the Operation in Syria Over Three Years]', *MK,* September 30, 2018, https://www.mk.ru/politics/2018/09/30/eksglavkom-vks-podvel-itogi-operacii-v-sirii-za-tri-goda.html.

Russia and VNSAs

Russian forces entered Syria following an official invitation from the Assad regime.[17] Russia's strategy was designed with a light footprint in mind given that a large deployment would not only be expensive and logistically unsound[18] but also politically unsustainable given a lack of domestic appetite for casualties. This casualty aversion was a product of both the Soviet intervention in Afghanistan in the 1980s[19] and the Chechen Wars of the 1990s.[20] The preferred option was to commit few Russian ground troops and then rely heavily on the use of the Russian Aerospace Forces (VKS hereafter) and with the real 'boots on the ground' being provided by VNSA proxy forces. As stated, these proxy forces were principally the YPG, and the Wagner Group.

Moscow's Use of the YPG in Syria

In 2012, the far-left Kurdish political party (with an armed wing), the *Partiya Karkerên Kurdistanê* (Kurdistan's Workers' Party or PKK), which operated in Turkey, reached an agreement with the Syrian government. According to this agreement, the PKK would take control of the Kurdish enclaves in the northern part of Syria along the border with Turkey. In return, the PKK – through its Syrian offshoot, the *Partiya Yekitiya Democrat* or Democratic Union Party (PYD hereafter) – would bring the then ongoing anti-Damascus protests in the Kurdish areas of Syria under control. It did this through the creation of various mechanisms such as regional committees and using the forces of the armed wing of the PYD, the YPG. The YPG was mentored and trained by the forces of the PKK.[21] These developments in 2012 were, of course, not welcomed by Turkey, which has been at war,

17 Manal, 'Syria's Ambassador to Russia Urges all Countries to Join Syria and Russia against Terrorism', *Syrian Arab News Agency*, October 1, 2015, https://sana.sy/en/?p=56454.

18 Eugeni Radugin, 'Putin o Siriyskoy Operacii: Trenirovka Dlya Rossiyskih Voysk [Putin On the Operation in Syria: Training of the Russian Forces]', *Politikus*, February 17, 2018, https://politikus.ru/events/104739-putin-o-siriyskoy-operacii-trenirovka-rossiyskih-voysk.html.

19 See Lester Grau and Michael A. Gress, *The Soviet-Afghan War: How a Superpower Fought and Lost*. (Kansas: Kansas University Press, 2002).

20 See Ariel Cohen, *Russia's Counterinsurgency in North Caucasus: Performance and Consequences*. (Carlisle, PA: US Army War College Press, 2014). https://publications.armywarcollege.edu/pubs/2261.pdf

21 Adam Baczko, Gilles Dorronsoro and Arthur Quesnay, 'The Kurds and the PKK' in *Civil War in Syria: Mobilization and Competing Social Orders*, ed. Adam Baczko, Gilles Dorronsoro and Arthur Quesnay (Cambridge: Cambridge University Press, 2017), 164-172.

with various levels of intensity, with the PKK within Turkey since 1984. Ankara, displeased with Assad and the freedom he had allowed the Syrian Kurds, had thus felt it necessary at that time to secure its southern border against the YPG. In addition, it threw its support behind both an anti-Assad Islamist group, Jabhat al-Nusra, and a fledgling VNSA that also stood in opposition to the Assad regime, the Free Syrian Army (FSA).[22] The FSA was also supported by the Western powers. Given that the Kurdish question was very much Turkey's Achilles' heel, ever since 2012 Ankara has come to oppose every actor in the Syrian conflict which tried to empower the YPG, including the US. Washington, for its part, did not perceive the YPG as being militarily linked to the PKK and itself used the YPG as a proxy.[23]

By September 2015, Moscow felt it was propitious to intervene itself in Syria as Assad began to look increasingly vulnerable to both Islamist groups and the FSA. But this intervention, as noted, had to be limited. Proxies were needed. The relationship with the YPG thus started at an early stage of Russia's engagement in Syria. While there exist different opinions as to why Moscow chose the YPG, including building on the historical relationship between the left-wing PKK and the Soviet Union[24] and the fact that they possessed knowledge of the area of operations, had linguistic skills and, most notably, a willingness to fight, the primary reason seems to have been more strategic. Using the YPG was very much in Russia's geopolitical interests. Firstly, making links with the YPG served 'to balance the US along the Euphrates line'.[25] Given Washington's use of the YPG, the Kremlin's apprehension was that the US would stay long-term in Syria and play the role of a 'guarantor' of Syrian-Kurd *de facto* independence. The creation of such a Kurdish quasi-state could, from Moscow's perspective, serve as a US bridgehead in the whole region. Additionally, such a quasi-state could

22 Lacin Idil Oztig, 'Syria and Turkey: Border-Security Priorities', *Middle East Policy*, 26:1 (Spring 2019), 120-123.

23 'Türkiye ve ABD arasındaki PYD krizi: Gün gün ne söylendi? [PYD Crisis Between Turkey and the US: Day to Day Developments]', *CNN Türk*, February 23, 2016. https://www.cnnturk.com/dunya/turkiye-ve-abd-arasindaki-pyd-krizi-gun-gun-ne-soylendi.

24 Gönül Tol, 'Why is Turkey Silent on Russia's Cooperation with the Syrian Kurds', *War on the Rocks*, December 20, 2017, https://warontherocks.com/2017/12/why-is-turkey-silent-on-russias-cooperation-with-the-syrian-kurds/; Dmitri Trenin, 'Putin's Plan for Syria: How Russia Wants to End the War', *Foreign Affairs*, December 13, 2017, https://www.foreignaffairs.com/articles/syria/2017-12-13/putins-plan-syria'.

25 Fehim Tastekin, 'Syrian Kurds Move Closer to Russia', *Al Monitor*, December 7, 2017, https://www.al-monitor.com/pulse/originals/2017/12/turkey-syria-kurds-move-closer-russia.html.

potentially threaten the territorial integrity of Syria and thus endanger the regime of Russia's ally, Assad.[26]

More important, though, for Moscow, was the effect that Russian support for the PYD/YPG would have on Turkey (a NATO member bordering Russia and anxious to see Assad removed). Influence with the YPG would mean influence with the PKK and thus leverage for Moscow in regard to Ankara.[27]

The decision to employ the YPG as a proxy was cemented by an incident in November 2015, when a Russian Su-24 fighter was shot down by Turkish forces.[28] Prior to this, Putin had been willing to turn a blind eye to Turkey's attempts to prevent the Syrian Kurds from creating a corridor between their enclaves of Kobani and Afrin in northern Syria. This, if successful, would have given the Kurds control of areas close to the Turkish border.[29] However, in the wake of this shootdown incident, there had to be, according to Russian political scientist, Sergei Markov, an 'asymmetric' response by Moscow to punish Ankara.[30] This response was to provide Russian support to Turkey's Syrian-Kurdish enemy, the PYD/YPG.[31] This newly formed Russia-PYD/YPG relationship was accentuated on several fronts through, for instance, the setting up in Moscow of the 'first European office' of the PYD[32] and by publicizing photos of the YPG

26 'Lavrov Zayavil O Popytkakh SSHA Sozdat' v Sirii Gosudarstvo s Oporoi na Kurdov [Lavrov Told About the US Attempt to Create a State with the Support of Kurds in Syria]', *Izvestiya*, February 16, 2018, https://iz.ru/709688/2018-02-16/lavrov-zaiavil-o-popytkakh-ssha-sozdat-v-sirii-gosudarstvo-s-oporoi-na-kurdov; Konstantin Sivkov, 'Siriiskiye Torgi [Syrian Bidding]', *Voenno-Promyshlenyi Kur'er*, 25:788 (July 2019), 1; 6.
27 Rod Thornton, 'Problems with the Kurds as Proxies Against Islamic State: Insights from the Siege of Kobane', *Small Wars & Insurgencies* 26, No. 6 (2015), 865-885.
28 'Incindent s Samoletom Su-24: Reakciya Rossiiskih Politikov [Incident with A Plane Su-24: Reaction of the Russian Politicians]', *TASS*, November 24, 2015, https://tass.ru/politika/2467278.
29 Nuri Elibol, 'Rus uçağını niçin düşürdük? [Why Did We Shoot Down the Russian Airplane?]', *Türkiye*, December 9, 2016, http://www.turkiyegazetesi.com.tr/yazarlar/nuri-elibol/594504.aspx. See also, Sami Kohen, 'Suriye'de kim kiminle, kime karşı? [Who with Whom and Against Whom in Syra?]', *Milliyet*, August 24, 2016, http://www.milliyet.com.tr/yazarlar/sami-kohen/suriye-de-kim-kiminle--kime-karsi--2300109/.
30 Sergei Markov, 'Nuzhen Assimetrichnyi Otvet [There Needs to Be an Asymmetric Response]', *Izvestia*, November 26, 2015, https://iz.ru/news/597265.
31 Markov, 'There Needs to be an Asymmetric Response'.
32 'Syria's Kurds to Open First European Office in Moscow', *Rudaw*, February 7, 2016, http://www.rudaw.net/english/middleeast/syria/070220161.

and Russian soldiers working together.[33] It all acted as a warning to Ankara: Turkey's enemy had Russian backing. As such, the PYD/YPG could be used as a 'stick' by Moscow; a coercive tool to use against Ankara which, as a consequence, would have to be more accommodating of Russian wishes.

The Russia-PYD/YPG 'cooperation' appeared to achieve the desired effect, certainly in terms of putting pressure on Ankara. For instance, Russia now managed to convince Turkey to help create 'de-escalation zones' in northern Syria. These were designed to encourage ceasefires and to allow space for political settlements to emerge (favourable, of course, to Assad and thus to Moscow). The groundwork for these was laid during the Astana Process (January through May 2017), which resulted in the setting up of five such zones in Idlib, Homs, eastern Ghouta, Deraa and Al-Quneitra. Russia, Turkey, and Iran were to serve as guarantors of these 'de-escalation. arrangements.[34]

Aside from its coercive/deterrent utility on the strategic level for Moscow vis-à-vis Turkey, the YPG also proved to be essential in combat against Russia's 'terrorist' opponents, in Syria. These were the IS and Jebhat Al-Nusra, operating mostly in the northern province of Aleppo.[35] Similarly, in Deir ez-Zor province, which, at the time of writing (October 2019), remains an important location for all the players in the conflict due to its oil resources as well as its geographic position (on the border between Syria and Iraq),[36] the Russians conducted joint operations with the YPG. Russian aircraft have carried out many sorties to support not just these joint operations but also independent activities by YPG forces. The YPG, in turn,

33 Ari Khalidi, 'Russian Soldiers' Pictures With Syrian Kurdish YPG Enrage Turkey', *Kurdistan24*, March 23, 2017, http://www.kurdistan24.net/en/news/b451ec1b-f144-46ea-8b65-e04da27b884f/russian-soldiers%E2%80%94pictures-with-syrian-kurdish-ypg-enrage-turkey; Tastekin, 'Syrian Kurds'.

34 'Final De-Escalation Zones Agreed on in Astana', *Al-Jazeera*, September 15, 2017, https://www.aljazeera.com/news/2017/09/final-de-escalation-zones-agreed-astana-170915102811730.html.

35 Aleksandr Dvornikov, 'Siria: Russkiy Grom [Syria: Russian Thunder]', interview by Yuri Gavrilov, *Rossiyskaya Gazeta – Federal'nyi Vypusk*, 6939.21, March 23, 2016, https://rg.ru/2016/03/23/aleksandr-dvornikov-dejstviia-rf-v-korne-perelomili-situaciiu-v-sirii.html.

36 Deir ez-Zor's area is populated by tribes which can be mobilized if necessary which makes it attractive for both Syria and Iran. See Alessandria Misa, 'Why Syria and the U.S. Clashed for Control East of the Euphrates', *Syria Deeply*, February 14, 2018, https://www.newsdeeply.com/syria/articles/2018/02/14/why-syria-and-the-u-s-clashed-for-control-east-of-the-euphrates; Sardar Mlla Drwish, 'Deir Ezzor: An Unbalanced Equation in the Syrian Conflict', *Atlantic Council*, February 21, 2018, http://www.atlanticcouncil.org/blogs/syriasource/deir-ezzor-an-unbalanced-equation-in-the-syrian-conflict.

has supplied intelligence to the Russians,[37] while also providing security for those Russian forces on the eastern bank of the Euphrates River.[38]

Turn in the Moscow-YPG Relationship

Proxies, however, and as the Russians have discovered, can be difficult to control. The YPG also had close ties with the US forces in Syria, having been incorporated into the Washington-backed Syrian Democratic Force (SDF) to fight the IS.[39] By 2018, the relationship between the YPG and Moscow had started to deteriorate as the links between the Syrian Kurds and the US forces became, as Markov put it, 'more advanced'.[40] As feared, the Syrian Kurds were forming a 'de-facto…pro-American puppet state on the territory of Syrian Kurdistan'.[41] Similar concerns were later expressed by the Russian Foreign Minister, Sergey Lavrov.[42] The Kurds' behaviour was at odds with Russia's dual aspirations to retain full control of one of the strongest VNSAs in Syria and to use the PYD/YPG as a tool of leverage over Ankara. Moscow's proxy was becoming too independent. There was a need thus for Moscow to show its authority over its YPG proxy.

The perfect opportunity for this arose with Turkey's decision in January 2018 to launch a cross-border intervention into the Kurdish-majority town of Afrin in northern Syria. In Operation *Olive Branch,* Ankara was seeking to oust from power in Afrin the perceived 'terrorists', that is, the YPG. The Russians, pointedly, stood aside.

---

37 'Plemya Izgo'ev: Mogut Li Kurdy Izmenit' Kartu Mira [The Rogue Tribe: Can the Kurds Change the World Map]', *Russia Today,* March 3, 2016, https://russian.rt.com/article/151653.
38 'Kurds to Provide Security for Russian Task Force on Eastern Bank of Euphrates River', *TASS,* December 4, 2017, http://tass.com/defense/978971; Nerdun Hacıoğlu, 'YPG, Russia Meeting 'Evaluated Steps to Take After ISIL': Russian Defense Ministry', *Hürriyet,* December 5, 2017, http://www.hurriyetdailynews.com/ypg-russia-meeting-evaluated-steps-to-take-after-isil-russian-defense-ministry-123603.
39 Thomas Gibbons-Neff, Jeremy White, and David Botti, 'The U.S. Has Troops in Syria. So Do the Russians and Iranians. Here's Where', *New York Times,* April 11, 2018. https://www.nytimes.com/interactive/2018/04/11/world/middleeast/syria-military-us-russia-iran.html
40 Kirill Vasil'yev, 'Eksperty: Tureckaya Voennaya Operaciya v Sirii Ne Privedet k Rostu Mezhdunarodnoi Napryazhennosti [Experts: Turkish Military Operation in Syria Won't Lead to an Increase of International Tension]', *Vechernyaya Moskva,* January 21, 2018, https://vm.ru/news/453694.html.
41 Vasil'yev, 'Turkish Military Operation'.
42 'Lavrov: SSHA Pytayutsya s Oporoi na Kurdov Sozdat' v Sirii Kvazigosudarstvo [Lavrov: The US is Trying to Create a Quasi-state in Syria with the Help of the Kurds]', *Vesti,* February 16, 2018, https://www.vesti.ru/doc.html?id=2988320&2988320.

Prior to *Olive Branch*, Russian forces had been cooperating with the YPG in Afrin to prevent a Turkish incursion.[43] Once the Turkish operation was underway, however, Russia redeployed its 170 military personnel to ensure their safety and refused to provide air support after a YPG request. Moscow, indeed, also prevented Damascus from using its own forces to shield the YPG from the Turkish assault (according to the Kurdish news outlet *Rudaw*).[44] Moreover, US forces could also not come to the Kurds' aid given that Afrin was seen to be squarely in the 'zone' of Russian control and US troops could not risk clashing with those of Russia or, indeed, with those of a NATO ally, Turkey.[45] The Kurds were left abandoned and exposed and they blamed Moscow. According to the YPG's commander, Sipan Hemo, the Russians had 'betrayed' the Kurds.[46]

It has been reasoned that Russian actions were aimed at demonstrating to the YPG the benefits of Moscow's 'friendship' and can be perceived as an act of warning, the aim being to ensure that the YPG did not go too far in terms of its ties with the West, namely, the US.[47] Moreover, as well as indicating to the Syrian Kurds just how much they needed Russian support, this Russian withdrawal from Afrin also helped the Kremlin to forge a closer bond with the Turkey's President Recep Tayyip Erdogan. Moscow now sought to influence Ankara through cooperation rather than leverage through coercion. According to the new Russian narrative (post the downing of the Su24), Turkey was as an important ally for Russia and, as such, Moscow would not stand in the way of Turkey's offensive in Afrin.[48]

Thus, the Kremlin's strategy in regard to Afrin served two particular purposes. Firstly, it showed that, when it came to the use of proxies, those

---

43 Umut Uras and Tamila Varshalomidze, 'What Is Russia's End Game in Afrin?' *Al-Jazeera*, January 23, 2018, https://www.aljazeera.com/indepth/features/russia-green-light-turkeys-afrin-offensive-180123064315810.html.

44 'Whedat hemayh alesh'eb: Rewseya khanetna wemn'et Demsheq men teqdeym aled'em lena fey mewajhh Terkeya [People's Protection Units: Russia Betrayed Us and Prevented Damascus from Supporting Us in The Fight Against Turkey]', *Rudaw*, January 22, 2018, http://www.rudaw.net/mobile/arabic/kurdistan/220120184.

45 'Russia Betrayed Us'.

46 'Russia Betrayed Us'.

47 Angelika Basisini, Alexandr Amasuncev and Georgii Makarenko, 'Dovody Moskvy: Pochemy Rossia ne vystupila protiv operacii Turcii v Sirii [Moscow's Arguments: Why Russia Did Not Protest Against the Turkish Operation in Syria]', *RBC*, January 22, 2018, https://www.rbc.ru/politics/22/01/2018/5a65be579a794762405e20d1?story=58c7ff469a7947398567fb3d.

48 Angelika Basisini, Alexandr Amasuncev and Georgii Makarenko, 'Dovody Moskvy: Pochemy Rossia ne vystupila protiv operacii Turcii v Sirii [Moscow's Arguments: Why Russia Did Not Protest Against the Turkish Operation in Syria]', *RBC*, January 22, 2018, https://www.rbc.ru/politics/22/01/2018/5a65be579a794762405e20d1?story=58c7ff469a7947398567fb3d.

that came within Moscow's orbit (such as the YPG) had to unequivocally accept Russian control. There was no sense of any quid pro quo loyalty. It was, in essence, a business relationship. And, moreover, when it came to power dynamics, Turkey was more important than the YPG. The Russian abandonment of the YPG at Afrin also served the second purpose: it helped Russia to improve its relationship with Turkey. It put the latter into a debtor position vis-à-vis Russia.[49] This debt was brought in later as Russian forces demanded that Turkey take action in regard to removing its support for the Islamist group, Jebhat Al-Nusra in Idlib province.[50]

Turkey, by mid-2018 and post-Afrin, could be more easily convinced by Moscow into removing support for Jebhat Al-Nusra, also known as Jebhat Fatah Al-Sham or Al Qaeda in Syria. Both Russia and the US had been displeased with Turkey's support for this offshoot of Al Qaeda. Moscow, though, was not just relying on soft-power influence here. It made sure that pressure on Ankara came from elsewhere as well. According to a Saudi political scientist, Yussef Diab, Assad's forces were making preparations, with Moscow's explicit support, to strike Al-Nusra in Idlib. As such, Erdogan was left with two options: to withdraw support from Al-Nusra and allow a Syrian/Russian free rein in Idlib or to continue supporting it and be labelled as the backer of a 'terrorist' organization in much of the Western world. To do the latter would not just strengthen Assad's hand internationally but also that of the PYD/YPG who *were* seen to be fighting 'terrorism'. They would both (Turkey's enemies) thus gain increased support from the international community, including from Russia. The YPG, especially, would be perceived as 'heroes' and Turkey would be left isolated.[51] Given these options, Ankara found itself between a rock and a hard place. However, accepting Russia's ultimatum to abandon Al Nusra seemed to be the more politically and militarily viable option. For Russia, it was certainly another victory. While the Afrin issue put the relationship between the Kremlin and the YPG in jeopardy, the former was able to use it not only to coerce/influence Turkey into accepting Moscow's wishes but it also assisted, to all intents and purposes, in achieving Russia's aim of neutralising Al-Nusra. And Russia did all this by the adroit playing of

---

49 Abdul Hamid Al-Madjali, "Eferyen 'ela teqat' aletreq! [Afrin at Crossroads!]', *ADDUSTOUR*, January 22, 2018, https://www.addustour.com/articles/994519.

50 Uras and Varshalomidze, 'What Is Russia's End Game in Afrin?'.

51 Aasma Bakhba, 'Deghewt Rewseya Weghedb Ameyreka Yedf'ean Terkeya Leltekhely 'en Alenserh [Russia's Pressure and American Anger Push Turkey to Abandon Al-Nusra]', *Al-Watan*, September 2, 2018, http://alwatan.com.sa/Politics/News_Detail.aspx?ArticleID=346532.

its YPG proxy as a specific tool to generate positive outcomes at a higher strategic level.

Overall, this close cooperation with the YPG served several purposes for Moscow. Firstly, the Kurds, given their skill set, acted as a significant force multiplier for a casualty-averse Russian military. The Russians have also used this group in order to both create leverage over Ankara and to balance the use of the YPG by US forces. It seems that Moscow has been very successful in using the YPG as a VSNA proxy. The US (having withdrawn) has not helped establish any pro-Western Syrian-Kurdish entity and Moscow appears to have developed very good relations now with Turkey, almost, it seems, to the point where it is pulling Turkey away from NATO.[52]

Russia's Use of a Private Military Company

In addition to its utilisation of a regional VNSA, Moscow has also made extensive use in Syria of a PMC known as the Wagner Group.[53] This is headed by Dmitri Valer'yevich Utkin, a former officer in Russian Military Intelligence (or GRU – nowadays just GU[54]) and consists (mostly) of Russian citizens who have served in the Russian armed forces. Despite the fact that PMCs remain technically illegal in Russia,[55] Wagner has been used semi-covertly to serve Russia's foreign policy objectives abroad, such as in Ukraine, Crimea, the Central African Republic, Libya, Sudan, Venezuela,

52 Ben Hubbard, Anton Troianovski, Carlotta Gall and Patrick Kingsley, 'In Syria, Russia is Pleased to Fill an American Void', *New York Times*, October 15, 2019, https://www.nytimes.com/2019/10/15/world/middleeast/kurds-syria-turkey.html.

53 Ilya Rozhdestvenskii, Anton Baev and Polina Rusyaeva, 'Prizraky Voyny: Kak V Sirii Poyavilas' Rossiyskaya Chastnaya Armiya [Ghosts of War: How the Russian Private Army Appeared in Syria]', *RBC*, August 25, 2016, https://www.rbc.ru/magazine/2016/09/57bac4309a79476d978e850d; 'Ih Prosto Net. Rassledovanie [They Simply Do Not Exist. Investigation]', *Novaya Gazeta*, October 9, 2017, https://www.novayagazeta.ru/articles/2017/10/09/74125-ih-prosto-net; Adam Taylor, 'What We Know About the Shadowy Russian Mercenary Firm Behind an Attack on U.S. Troops in Syria', *Washington Post*, February 23, 2018, https://www.washingtonpost.com/news/worldviews/wp/2018/02/23/what-we-know-about-the-shadowy-russian-mercenary-firm-behind-the-attack-on-u-s-troops-in-syria; 'Syria War: Who are Russia's Shadowy Wagner Mercenaries?' *BBC*, February 23, 2018, https://www.bbc.com/news/world-europe-43167697. Wagner was not the only PMC used by Russia. The other one was *Shchit* (Shield). See Denis Korotkov, 'Bez Shchita [Without a Shield]', *Novaya Gazeta*, June 28, 2019, https://www.novayagazeta.ru/articles/2019/07/28/81406-bez-schita.

54 GU stands for General Directorate of the General Staff of the Armed Forces of the Russian Federation or *Glavnoe Upravlenie General'nogo Shtaba Voennykh Sil Rossiiskoi Federacii.*

55 Igor Pushkarev, 'Kak Vyglyadit Lager' ChVK Wagnera v Krasnodare [What the Camp of the Wagner PMC Looks like in Krasnodar]', *ZNAK*, March 5, 2018, https://www.znak.com/2018-03-05/kak_vyglyadit_lager_chvk_vagnera_v_krasnodare.

and, most notably, in Syria.[56] It remains unclear whether or not the group is controlled by the GU, that is, by the Ministry of Defence. Some Russian and foreign sources indicate that Wagner is actually controlled by the civilian Russian intelligence service, the Federal Security Service (*Federal'naya Sluzhba Bezopasnosti* or FSB).[57]

The underlying reason for such outsourcing to this PMC could be mainly attributed to the political and strategic advantages offered. As Peter Singer notes, '…the deaths of private contractors may be less politically sensitive than those of soldiers'.[58] And this is certainly the case with the Kremlin: it is casualty conscious. Casualties in Syria had to be kept to a minimum lest Russia lose domestic support for its intervention. The overall image of the campaign had to depict Russia's success but without major losses.[59] Moscow could utilise Syrian government forces and those of the YPG but their use was not ideal. They each could not be used, for instance, in certain areas (given Arab/Kurdish antipathy) and their reliability was always open to question. The use thus of the more 'professional' Wagner forces provided for greater chance of success in any particular mission. Also, because they were not official Russian troops they provided 'plausible deniability' if ever any operations went awry or involved questionable or unethical activities.[60] Such 'tactical ambiguity' complicated any political choices and, consequently, the practice of strategy for other participants due to the creation of a smoke-screen, thereby giving Moscow more flexibility in its actions.[61]

---

56 Muhammad Fadl Ali, 'Alemretzeqh Alerwes Fey Shewar' Alekhertewm Aleywem Nefteqd Alesheyw'eyh Aledwelyh [Russian Mercenaries on the Streets of Khourtum. Today We Miss International Communism]', *Sudanese Online*, January 10, 2019, https://sudaneseonline.com/board/7/msg/1547173785.html; Maria Tsvetkova and Anton Zverev, 'Exclusive: Kremlin-led Contractors Help Guard Venezuela's Maduro – Sources', *Reuters*, January 25, 2019, https://www.reuters.com/article/us-venezuela-politics/u-s-steps-up-pressure-on-maduro-as-russia-backs-venezuelan-ally-idUSKCN1PJ1QR; 'Mertezqh Rews Leqm' Ahetjajat Aleswedan. Werwesya Walebsheyr Yenfeyan [Russian Mercenaries to Suppress Sudan Protests…Russia and Bashir Deny]', *Noonpost*, January 24, 2019, http://www.noonpost.com/content/26308.

57 Rozhdestvenskii, Baev, Rusyaeva, 'Ghosts of War'; Mihnad Al-Haj Ali, ''En Alasetkhebarat Alerwesyh Fey Sewreya [About Russian Intelligence in Syria]', *Al Modon*, October 25, 2018, https://www.almodon.com/opinion/2018/10/22/.

58 Peter W. Singer, *Corporate Warriors: The Rise of the Privatized Military Industry*. (Ithaca: Cornell University Press, 2007), 4.

59 Indeed, Putin was re-elected in March 2018.

60 Roger Boyes, 'Putin's Private Army in a War with No Rules', *The Times*, November 16, 2019, https://www.thetimes.co.uk/article/putin-s-private-army-is-fighting-a-war-with-no-rules-c092vzhcd; Allison Stranger and Marc E. Williams, 'Private Military Corporations: Benefits and Costs of Outsourcing Security', *Yale Journal of International Affairs*, 2:1 (2006) 8; Rozhdestvenskii, Baev, Rusyaeva, 'Ghosts of War'.

61 Lukas Milevski, 'The Strategic Response to Ambiguity', *Orbis* 63, no. 3 (2019): 378-379; 382.

According to an anonymous source from the Russian MoD, the Wagner Group first entered the Middle Eastern theatre shortly before Russia's official deployment to Syria in fall 2015.[62] Some 2,500 Wagner personnel came to be stationed around Latakia and Aleppo.[63] Although, by 2016 there were said to be only between 1,000 and 1,600 Wagner troops operating in Syria. Such information, though, was never confirmed by the MoD.[64]

This level of commitment by Wagner was important in terms of achieving military effectiveness without making their presence obvious. In terms of their missions, the first major one, according to a former employee, was in a key role they played in retaking Palmyra in March 2016 from the IS. He noted that Wagner forces (who lost 32)[65] went in first, followed by Russian ground forces and then Syrian government troops. The primary task of this PMC in Syria, according to this same source, was to advance in 'difficult' regions in order to minimize the casualties of both Russian and Syrian troops.[66]

In October 2017, the Wagner Group also fought at Deir ez-Zor, helping to free the city from IS control. Their presence was confirmed by several sources. In an interview with the Russian newspaper *Svobodnaya Pressa*, Kerem Yildirim, a Turkish political scientist, stated that YPG forces were taken by surprise when they reached Abu Kamal, a city in Deir ez-Zor Governorate and clashed with what he called 'Russian military contractors',[67] presumably from the Wagner Group.

In the same month, when the IS launched a counterattack in Idlib, Wagner's reconnaissance-assault company was deployed there and then later to the highway between Palmyra and Deir-ez-Zor to help halt the attacks. According to *Novaya Gazeta*, an engagement between IS fighters and Syrian government forces supported by Russian airpower, Russian Special Forces and by Wagner personnel resulted in the deaths of several Russian military personnel and Wagner members. The Russian MoD, perhaps not surprisingly, while registering the casualties of the

---

62 Rozhdestvenskii, Baev, Rusyaeva, 'Ghosts of War'.
63 Rozhdestvenskii, Baev, Rusyaeva, 'Ghosts of War'; Thomas Grove, 'Up To Nine Russian Contractors Die in Syria, Experts Say', *Wall Street Journal*, December 18, 2015, https://www.wsj.com/articles/up-to-nine-russian-contractors-die-in-syria-experts-say-1450467757.
64 Rozhdestvenskii, Baev, Rusyaeva, 'Ghosts of War'.
65 Boyes, 'Putin's Private Army'.
66 Rozhdestvenskii, Baev, Rusyaeva, 'Ghosts of War'.
67 Zaur Karaev, 'Kurdy Stolknulis' S Rossiyskimi Nayemnikami [The Kurds Clashed with Russian Contractors]', *Svobodnaya Pressa*, November 27, 2017, http://svpressa.ru/war21/article/187161/.

regular forces, released no official statement regarding the deaths of the contractors.[68] In essence, these contractors were akin to mercenaries and still 'illegal'.[69] Their deaths, just as with their actual presence, could not be announced, they were 'ghosts'.[70] Some even argue that Wagner troops can be classified as 'disposable'.[71] Here is the dilemma for any state employing military contractors; they should not be used in combat situations because, not being state employees, their status under international law is murky at best.[72]

There was another major Wagner operation a few months later, in February 2018. But this time the number of Wagner deaths could not readily be ignored. The Kremlin was seemingly taken by surprise by Wagner's involvement in an advance near Deir ez-Zor in which, according to estimates, hundreds of Wagner contractors were killed in an airstrike by the US Coalition Forces.[73] The official Russian MoD version only acknowledged the presence of Syrian militia forces at Deir ez-Zor, stressing that the clash took place due to the fact that these forces had not coordinated their operation with the commander of the Russian task force and that there were no Russian military personnel in the region.[74] The official story stressed that it was the Syrian militia (not government troops) who suffered the casualties and not Wagner. Had there indeed been an acknowledged clash between Russian Wagner and US forces, the outcome could, in the worst-case scenario, have, as one source puts it, '...plunge[d] both countries into bloody conflict'.[75]

---

68 'They Simply Do Not Exist'. The article provided by *Novaya Gazeta* is, according to the newspaper, based on interviews with former Wagner employees and MoD officials whose names are not disclosed.

69 On the legality debate of PMCs in Russia, see Kimberly Marten, 'Russia's Use of Semi-State Security Force: The Case of the Wagner Group', *Post-Soviet Affairs* 35, no. 3 (2019): 184-188.

70 Rozhdestvenskii, Baev, Rusyaeva, 'Ghosts of War'.

71 Milevski, 'The Strategic Response to Ambiguity', 387-388.

72 Lindsay Cameron, 'Private Military Companies Their Status under International Humanitarian Law and Its Impact on Their Regulation', *International Review of the Red Cross,* 88:863 (2006): 582.

73 'CIT Nazvala Pervye Imena Rossiyskih Naemnikov, Pogibshih v Sirii 7 Fevralya [CIT Named the First Names of the Russian Contractors Who Died in Syria on February 7]', *ZNAK,* February 12, 2018, https://www.znak.com/2018-02-12/cit_nazvala_pervye_imena_rossiyskih_naemnikov_pogibshih_v_sirii_7_fevralya.

74 'Popavshie Pod Udar Koalicii SSHA Siriiskiye Opolchency Ne Soglasovali Operaciyu s RF [Syrian Militias Who Were Hit by the US Coalition Have Not Coordinated the Operation with the Russian Federation]', *Interfax,* February 8, 2018, https://www.interfax.ru/world/599096.

75 Gibbons-Neff et al., 'The U.S. Has Troops in Syria'.

It appears that Wagner's attempt to seize the gas fields in Deir ez-Zor was without the Kremlin's knowledge or approval.[76] Here is another risk with using contractors as proxies. Being outside state control and being businesses, they will sometimes seek business opportunities. This, it certainly appears, is what happened at Deir ez-Zor. Some 500 Wagner Group employees, along with their Syrian militia allies, engaged in some 'private enterprise' in attempting to seize a gas plant protected by US Special Forces.[77] Hence the use of Coalition airpower to halt their advance.[78]

It is interesting here to note the quality of the Wagner troops. As one US analyst put it, 'It took America's most elite troops and advanced aircraft to defeat 500 mercenaries. What happens when they [US forces] have to face a thousand? Five thousand? More?'[79]

Overall, the use of Wagner in Syria presented a viable alternative to deploying Russian troops. The Kremlin tried to deny their deployment despite the fact that on several occasions, information relating to the group's presence and activities in Syria had reached the public domain. While the multitude of articles that were published by both Russian and international newspapers were damaging from the public relations perspective,[80] this did not outweigh the strategic and military advantages of using this PMC in Syria. However, what remains of concern for the Kremlin is the actual control of the Wagner Group. Due to its degree of autonomy, and presumable lack of integration into the Russian task force command structure that was evident in Syria, it is difficult to assess to what degree such PMCs could be integrated in future Russian overall military operations.

76 Neil Hauer, 'Russia's Mercenary Debacle in Syria', *Foreign Affairs*, February 26, 2018, https://www.foreignaffairs.com/articles/syria/2018-02-26/russias-mercenary-debacle-syria; Maxim Solopov, Maxim Yusin, Mariana Belen'kaya, Valeria Mishina, Yelena Chernenko and Ivan Safronov, 'Chastniki Boevyh Deystviy: Diplomaty Utochnayut, Kto Pogib Ot Amerikanskogo Udara [Private Contractors of the Fighting: Diplomats Clarify Who Died from the US Strike]', *Kommersant*, 27 (February 2018) 3.

77 Hauer, 'Russia's Mercenary Debacle'.

78 Boyes, 'Putin's Private Army'.

79 Professor Sean McFate quoted in Boyes, 'Putin's Private Army'.

80 See, for instance, '"L'enh jeysh faghenr" telaheq alesa'eyen ela feth melfat alemretzeqh fey sewreya ['The Curse of the Wagner Army' – Hunt for Those Seeking to Uncover the Mercenary Case in Syria]', *Aawsaat*, January 2, 2019, https://aawsat.com/home/article/1528716.

Conclusion

In conclusion, in its Syrian campaign Moscow brought into play two specific types of VNSA as part of its overall strategy. Firstly, it used an indigenous non-state actor, the YPG of the Syrian Kurds. The YPG were very effective fighters and provided Russian forces with valuable intelligence. They were also useful in that manipulating them could also serve the wider need to establish a degree of influence over Turkish actions in Syria. Moreover, by establishing a close relationship with the YPG Moscow could also undermine supposed US ambitions in the region. Indeed, it was a relationship that could also undermine NATO itself by creating splits between Ankara and Washington. Russia's use of the YPG thus proved to be beneficial on the military, strategic and geopolitical levels.

Secondly, Russia heavily relied on a PMC, the Wagner Group. Unlike in the case of the YPG, Wagner's purpose was primarily military in nature. It has had distinct advantages. Wagner forces could be used anywhere in Syria, unlike the YPG which was mostly tied to the Kurdish areas and also could not operate alongside Assad's Arab government forces. Wagner provided experienced troops who were also, to a degree, 'expendable' and their actions could always be 'plausibly deniable' which would help avoid strategic and political consequences. Wagner seems to have constituted a pivotal element in Russia's military strategy from the very onset of the campaign. However, the integration of its forces came to raise concerns in that they occasionally were out of Russian state control. All this being said, Wagner's activities appear to have been vital in Russia emerging as the undoubted 'victor' – the 'big winner' – in the whole Syrian imbroglio.

In its utilisation of both the YPG and Wagner, Moscow, despite a few missteps, can be said to have handled both with some dexterity and to have produced some very positive results.

References

Al-Haj Ali, M. 'En Alasetkhebarat Alerwesyh Fey Sewreya [About Russian Intelligence in Syria]'. *Al Modon*, October 25, 2018. https://www.almodon.com/opinion/2018/10/22/.

Al-Madjali, A.H. 'Eferyen "ela teqat" aletreq! [Afrin at Crossroads!]'. *ADDUSTOUR*, January 22, 2018. https://www.addustour.com/articles/994519.

Associated Press. 'Why Russia is the Big Winner from Trump's Syria Pullback and Ukraine Turmoil'. *Haaretz*, October 13, 2019. https://www.haaretz.com/

us-news/why-russia-is-the-big-winner-from-trump-s-syria-pullback-and-ukraine-turmoil-1.7970597.

Baczko, A., G. Dorronsoro and A. Quesnay. 'The Kurds and the PKK'. *Civil War in Syria: Mobilization and Competing Social Orders*, edited by Adam Baczko, Gilles Dorronsoro and Arthur Quesnay, Cambridge: Cambridge University Press, 2017. https://doi.org/10.1017/9781108355322.010.

Bakhba, A. 'Deghewt Rewseya Weghedb Ameyreka Yedf'ean Terkeya Leltekhely 'en Alenserh [Russia's Pressure and American Anger Push Turkey to Abandon Al-Nusra]'. *Al-Watan*, September 2, 2018. http://alwatan.com.sa/Politics/News_Detail.aspx?ArticleID=346532.

Barshad, A. 'Putin's Rasputin'. *Longread*, July 11, 2019. https://longreads.com/2019/07/11/putins-rasputin/.

Basisini, A., A. Amasuncev, and G. Makarenko. 'Dovody Moskvy: Pochemy Rossia ne vystupila protiv operacii Turcii v Sirii [Moscow's Arguments: Why Russia Did Not Protest Against the Turkish Operation in Syria]'. RBC, January 22, 2018. https://www.rbc.ru/politics/22/01/2018/5a65be579a794762405e20d1?story=58c7ff469a7947398567fb3d.

Boyes, R. 'Putin's Private Army in a War with No Rules'. *The Times*, November 16, 2019. https://www.thetimes.co.uk/article/putin-s-private-army-is-fighting-a-war-with-no-rules-c092vzhcd.

Bunker, R.J. 'Defeating Violent Nonstate Actors'. *Parameters*, 43:4 (2013) 57-65.

Cameron, L. 'Private Military Companies Their Status under International Humanitarian Law and Its Impact on Their Regulation'. *International Review of the Red Cross*, 88:863 (2006) 573-598.

Charap, S. 'Russia, Syria and the Doctrine of Intervention'. *Survival*, 55:1 (2013) 35-41.

'CIT Nazvala Pervye Imena Rossiyskih Naemnikov, Pogibshih v Sirii 7 Fevralya [CIT Named the First Names of the Russian Contractors Who Died in Syria on February 7]'. ZNAK, February 12, 2018. https://www.znak.com/2018-02-12/cit_nazvala_pervye_imena_rossiyskih_naemnikov_pogibshih_v_sirii_7_fevralya.

Cohen, A. *Russia's Counterinsurgency in North Caucasus: Performance and Consequences*. Carlisle, PA: US Army War College Press, 2014.

Dasgupta, S. *Paramilitary Forces and Security Reorganization*. GSC Quarterly. Program on Global Security and Cooperation, Social Science Research Council (2004). http://www.ssrc.org.

Drwish, S.M. 'Deir Ezzor: An Unbalanced Equation in the Syrian Conflict'. *Atlantic Council*, February 21, 2018, http://www.atlanticcouncil.org/blogs/syriasource/deir-ezzor-an-unbalanced-equation-in-the-syrian-conflict.

Dugin, A. *Osnovy Geopolitiki: Geopoliticheskoe Budushhee Rossii* [The Foundations of Geopolitics: Russia's Geopolitical Future]. Moscow: Arktogeja, 1997.

Dvornikov, A. 'Siria: Russkiy Grom [Syria: Russian Thunder]'. Interview by Yuri Gavrilov, *Rossiyskaya Gazeta – Federal'nyi Vypusk*, 6939.21, March 23, 2016. https://rg.ru/2016/03/23/aleksandr-dvornikov-dejstviia-rf-v-korne-perelomili-situaciiu-v-sirii.html.

Eizenstat, S. E.; Porter, Stu John Edward; and Weinstein, Jeremy M. 'Rebuilding Weak States'. *Foreign Affairs*, 84:1 (2005) 134-146.

Elibol, N. 'Rus uçağını niçin düşürdük? [Why Did We Shoot Down the Russian Airplane?]'. *Türkiye*, December 9, 2016. http://www.turkiyegazetesi.com.tr/yazarlar/nuri-elibol/594504.aspx.

Fadl Ali, M. 'Alemretzeqh Alerwes Fey Shewar' Alekhertewm Aleywem Nefteqd Alesheyw'eyh Aledwelyh [Russian Mercenaries on the Streets of Khourtum. Today We Miss International Communism]'. *Sudanese Online*, January 10, 2019. https://sudaneseonline.com/board/7/msg/1547173785.html.

'Final De-Escalation Zones Agreed on in Astana'. *Al-Jazeera*, September 15, 2017. https://www.aljazeera.com/news/2017/09/final-de-escalation-zones-agreed-astana-170915102811730.html.

Gibbons-Neff, T., J. White,and D. Botti. 'The U.S. Has Troops in Syria. So Do the Russians and Iranians. Here's Where'. *New York Times*, April 11, 2018. https://www.nytimes.com/interactive/2018/04/11/world/middleeast/syria-military-us-russia-iran.html.

Grau, L. and M.A. Gress. *The Soviet-Afghan War: How a Superpower Fought and Lost*. Kansas: Kansas University Press, 2002.

Grove, T. 'Up to Nine Russian Contractors Die in Syria, Experts Say'. *Wall Street Journal*, December 18, 2015. https://www.wsj.com/articles/up-to-nine-russian-contractors-die-in-syria-experts-say-1450467757.

Grynkewich, A. G. 'Welfare as Warfare: How Violent Non-State Groups Use Social Services to Attack the State'. *Studies in Conflict & Terrorism*, 31:4 (2008) 350-370.

Hacıoğlu, N. 'YPG, Russia Meeting 'Evaluated Steps to Take After ISIL': Russian Defense Ministry'. *Hürriyet*, December 5, 2017. http://www.hurriyetdailynews.com/ypg-russia-meeting-evaluated-steps-to-take-after-isil-russian-defense-ministry-123603.

Hauer, N. 'Russia's Mercenary Debacle in Syria'. *Foreign Affairs*, February 26, 2018. https://www.foreignaffairs.com/articles/syria/2018-02-26/russias-mercenary-debacle-syria.

Hubbard, B., A. Troianovski, C. Gall and P. Kingsley. 'In Syria, Russia is Pleased to Fill an American Void'. *New York Times*, October 15, 2019. https://www.nytimes.com/2019/10/15/world/middleeast/kurds-syria-turkey.html.

Jackson, R. H. *Quasi-States: Sovereignty, International Relations, and the Third World*. Cambridge: Cambridge University Press, 1990.

Karaev, Z. 'Kurdy Stolknulis' S Rossiyskimi Nayemnikami [The Kurds Clashed with Russian Contractors]'. *Svobodnaya Pressa*, November 27, 2017. http://svpressa.ru/war21/article/187161/.

Khalidi, A. 'Russian Soldiers' Pictures with Syrian Kurdish YPG Enrage Turkey'. *Kurdistan24*, March 23, 2017. http://www.kurdistan24.net/en/news/b451ec1b-f144-46ea-8b65-e04da27b884f/russian-soldiers%E2%80%94pictures-with-syrian-kurdish-ypg-enrage-turkey.

Kohen, S. 'Suriye'de kim kiminle, kime karşı?' [Who with Whom and Against Whom in Syra?]'. *Milliyet*, August 24, 2016. http://www.milliyet.com.tr/yazarlar/sami-kohen/suriye-de-kim-kiminle--kime-karsi--2300109/.

Korotkov, D. 'Bez Shchita [Without a Shield]'. *Novaya Gazeta,* June 28, 2019. https://www.novayagazeta.ru/articles/2019/07/28/81406-bez-schita.

Krechetnikov, A. 'Medvedev y Putin: Govorim Liviya, Podrozumevayem Kreml' [Medvedev and Putin: Saying Libya, Thinking Kremlin]'. *BBC,* March 22, 2011. https://www.bbc.com/russian/russia/2011/03/110322_medvedev_putin_libya.

'Kurds to Provide Security for Russian Task Force on Eastern Bank of Euphrates River'. *TASS,* December 4, 2017. http://tass.com/defense/978971.

'Ih Prosto Net. Rassledovanie [They Simply Do Not Exist. Investigation]'. *Novaya Gazeta,* October 9, 2017. http://www.novayagazeta.ru/articles/2017/10/09/74125-ih-prosto-net.

'Incident s Samoletom Su-24: Reakciya Rossiiskih Politikov [Incident with A Plane Su-24: Reaction of the Russian Politicians]'. *TASS,* November 24, 2015. https://tass.ru/politika/2467278.

'"L'enh jeysh faghenr" telaheq alesa'eyen ela feth melfat alemretzeqh fey sewreya ['The Curse of the Wagner Army' – Hunt for Those Seeking to Uncover the Mercenary Case in Syria]'. *Aawsaat,* January 2, 2019. https://aawsat.com/home/article/1528716/.

'Lavrov: SSHA Pytayutsya s Oporoi na Kurdov Sozdat' v Sirii Kvazigosudarstvo [Lavrov: The US is Trying to Create a Quasi-state in Syria with the Help of the Kurds]'. *Vesti,* February 16, 2018. https://www.vesti.ru/doc.html?id=2988320&2988320.

'Lavrov Zayavil O Popytkakh SSHA Sozdat' v Sirii Gosudarstvo s Oporoi na Kurdov [Lavrov Told About the US Attempt to Create a State with the Support of Kurds in Syria]'. *Izvestiya,* February 16, 2018, https://iz.ru/709688/2018-02-16/lavrov-zaiavil-o-popytkakh-ssha-sozdat-v-sirii-gosudarstvo-s-oporoi-na-kurdov.

Mackintosh, J. M. *Strategy and Tactics of Soviet Foreign Policy.* New York: Oxford University Press, 1963.

Markov, S. 'Nuzhen Assimetrichnyi Otvet [There Needs to Be an Asymmetric Response]'. *Izvestia,* November 26, 2015. https://iz.ru/news/597265.

Marten, K. 'Russia's Use of Semi-State Security Forces: The Case of the Wagner Group'. *Post-Soviet Affairs* 35, no. 3 (2019): 181-204. https://doi.org/10.1080/1060586X.2019.1591142.

'Mertezqh Rews Leqm' Ahetjajat Aleswedan. Werwesya Walebsheyr Yenfeyan [Russian Mercenaries to Suppress Sudan Protests…Russia and Bashir Deny]'. Noonpost, January 24, 2019. http://www.noonpost.com/content/26308.

Milevski, L. 'The Strategic Response to Ambiguity'. *Orbis* 63, no 3 (Spring 2019): 376-390. https://doi.org/10.1016/j.orbis.2019.05.001.

Misa, A. 'Why Syria And the U.S. Clashed for Control East of The Euphrates'. *Syria Deeply,* February 14, 2018. https://www.newsdeeply.com/syria/articles/2018/02/14/why-syria-and-the-u-s-clashed-for-control-east-of-the-euphrates.

Mumford, A. *Proxy Warfare.* Cambridge, UK: Polity Press, 2013.

Oztig, L.I. 'Syria and Turkey: Border-Security Priorities'. *Middle East Policy* 26, no. 1 (Spring 2019): 117-126. https://doi.org/10.1111/mepo.12404.

'Plemya Izgo'ev: Mogut Li Kurdy Izmenit' Kartu Mira [The Rogue Tribe: Can the Kurds Change the World Map]'. *Russia Today*, March 3, 2016. https://russian.rt.com/article/151653.

'Popavshie Pod Udar Koalicii SSHA Siriiskiye Opolchency Ne Soglasovali Operaciyu s RF [Syrian Militias Who Were Hit by the US Coalition Have Not Coordinated the Operation with the Russian Federation]'. *Interfax*, February 8, 2018. https://www.interfax.ru/world/599096.

'Punkt Material'no-Tekhnicheskogo Obezpecheniya VMF Rossii v Tartuse. Dos'ye [Logistical Point of the Russian Navy in Tartus. Dossier]'. *TASS*, December 13, 2017. https://tass.ru/info/4808523.

Pushkarev, I. 'Kak Vyglyadit Lager' ChVK Wagnera v Krasnodare [What the Camp of the Wagner PMC Looks like in Krasnodar]'. *ZNAK*, March 5, 2018. https://www.znak.com/2018-03-05/kak_vyglyadit_lager_chvk_vagnera_v_krasnodare.

'Putin Nazval Osnovnuyu Zadachu Rossiyskih Voennyh v Sirii [Putin Named the Main Aim for the Russian Military in Syria]'. *Interfax*, October 11, 2015. https://www.interfax.ru/russia/472593.

Putin, V. 'Poslaniye Federal'nomy Sobraniyu Rossiiskoi Federacii [Message to the Federal Assembly of the Russian Federation]'. *Kremlin*, April 25, 2005. http://kremlin.ru/events/president/transcripts/22931.

Radugin, E. 'Putin o Siriyskoy Operacii: Trenirovka Dlya Rossiyskih Voysk [Putin On the Operation in Syria: Training of the Russian Forces]'. *Politikus*, February 17, 2018. https://politikus.ru/events/104739-putin-o-siriyskoy-operacii-trenirovka-rossiyskih-voysk.html.

Roche, C. 'Syrian Opposition Factions in the Syrian Civil War'. *Bellingcat*, August 13, 2016. https://www.bellingcat.com/news/mena/2016/08/13/syrian-opposition-factions-in-the-syrian-civil-war/.

Rozhdestvenskii, I., A. Baev and P. Rusyaeva. 'Prizraky Voyny: Kak V Sirii Poyavilas' Rossiyskaya Chastnaya Armiya [Ghosts of War: How the Russian Private Army Appeared in Syria]'. *RBC*, August 25, 2016. https://www.rbc.ru/magazine/2016/09/57bac4309a79476d978e850d.

Singer, P. W. Corporate Warriors: The Rise of the Privatized Military Industry. Ithaca: Cornell University Press, 2007.

Sivkov, K. 'Siriiskiye Torgi [Syrian Bidding]'. *Voenno-Promyshlenyi Kur'er*, 25:788 (July 2019): 1; 6.

Solopov, M., M. Yusin, M. Belen'kaya, V. Mishina, V. Chernenko and I. Safronov. 'Chastniki Boevyh Deystviy: Diplomaty Utochnayut, Kto Pogib Ot Amerikanskogo Udara [Private Contractors of the Fighting: Diplomats Clarify Who Died from the US Strike]'. *Kommersant*, 27 (February 2018).

Stranger, A. and M.E. 'Private Military Corporations: Benefits and Costs of Outsourcing Security'. *Yale Journal of International Affairs*, 2:1 (2006) 4-19.

'Syria War: Who are Russia's Shadowy Wagner Mercenaries?' *BBC*, February 23, 2018. https://www.bbc.com/news/world-europe-43167697.

'Syria's Ambassador to Russia Urges all Countries to Join Syria and Russia against Terrorism'. *Syrian Arab News Agency*, October 1, 2015. https://sana.sy/en/?p=56454.

'Syria's Kurds to Open First European Office in Moscow'. *Rudaw*, February 7, 2016. http://www.rudaw.net/english/middleeast/syria/070220161.

Tastekin, F. 'Syrian Kurds Move Closer to Russia'. *Al Monitor*, December 7, 2017. https://www.al-monitor.com/pulse/originals/2017/12/turkey-syria-kurds-move-closer-russia.html.

Taylor, A. 'What We Know About the Shadowy Russian Mercenary Firm Behind an Attack on U.S. Troops in Syria'. *Washington Post*, February 23, 2018. https://www.washingtonpost.com/news/worldviews/wp/2018/02/23/what-we-know-about-the-shadowy-russian-mercenary-firm-behind-the-attack-on-u-s-troops-in-syria.

Thomas, T. S. and W.D. Casebeer. *Violent Systems: Defeating Terrorists, Insurgents, and Other Non-State Adversaries*. Colorado Springs: USAF Institute for National Security Studies, 2004.

Thornton R. 'Problems with the Kurds as Proxies Against Islamic State: Insights from the Siege of Kobane'. *Small Wars & Insurgencies*, 26:6 (2015) 865-885. https://doi.org/10.1080/09592318.2015.1095844.

Thornton, R. 'The Russian Military's New "Main Emphasis"'. *The RUSI Journal*, 162:4 (2017) 18-28. https://doi.org/10.1080/03071847.2017.1381401.

Tol, G. 'Why is Turkey Silent on Russia's Cooperation with the Syrian Kurds'. *War on the Rocks*, December 20, 2017. https://warontherocks.com/2017/12/why-is-turkey-silent-on-russias-cooperation-with-the-syrian-kurds/.

Trenin, D. 'Putin's Plan for Syria: How Russia Wants to End the War'. *Foreign Affairs*, December 13, 2017. https://www.foreignaffairs.com/articles/syria/2017-12-13/putins-plan-syria'.

'Trump Orders US Troops Out of Northern Syria as Turkish Assault Continues'. *The Guardian*, October 13, 2019. https://www.theguardian.com/world/2019/oct/13/trump-us-troops-northern-syria-turkish-assault-kurds.

Tsvetkova, M. and A. Zverev. 'Exclusive: Kremlin-led Contractors Help Guard Venezuela's Maduro – Sources'. *Reuters*, January 25, 2019, https://www.reuters.com/article/us-venezuela-politics/u-s-steps-up-pressure-on-maduro-as-russia-backs-venezuelan-ally-idUSKCN1PJ1QR.

'Türkiye ve ABD arasındaki PYD krizi: Gün gün ne söylendi? [PYD Crisis Between Turkey and the US: Day to Day Developments]'. CNN Türk, February 23, 2016. https://www.cnnturk.com/dunya/turkiye-ve-abd-arasindaki-pyd-krizi-gun-gun-ne-soylendi.

Uras, U. and T. Varshalomidze. 'What Is Russia's End Game in Afrin?' *Al-Jazeera*, January 23, 2018. https://www.aljazeera.com/indepth/features/russia-green-light-turkeys-afrin-offensive-180123064315810.html.

Val'chenko, S. 'Eks-Glavkom VKS Podvel Itogi Operacii v Sirii za Tri Goda [Former Commander-in-Chief of the VKS Summed Up the Results of the Operation in Syria Over Three Years]'. *MK*, September 30, 2018. https://www.mk.ru/politics/2018/09/30/eksglavkom-vks-podvel-itogi-operacii-v-sirii-za-tri-goda.html.

Vasil'yev, K. 'Eksperty: Tureckaya Voennaya Operaciya v Sirii Ne Privedet k Rostu Mezhdunarodnoi Napryazhennosti [Experts: Turkish Military Operation in

Syria Won't Lead to an Increase of International Tension]'. *Vechernyaya Moskva,* January 21, 2018. https://vm.ru/news/453694.html.

'Voyna v Sirii Ne Zakonchena, Zayavil Lavrov [The War in Syria Has Not Ended, Said Lavrov]'. *RIA,* April 3, 2019. https://ria.ru/20190403/1552359037.html.

'Whedat hemayh alesh'eb: Rewseya khanetna wemn'et Demsheq men teqdeym aled'em lena fey mewajhh Terkeya [People's Protection Units: Russia Betrayed Us and Prevented Damascus from Supporting Us in The Fight Against Turkey]'. *Rudaw,* January 22, 2018. http://www.rudaw.net/mobile/arabic/kurdistan/220120184.

7

UNEASY ALLIANCES

The Political Consequences of Arming Non-State Groups in Counter-Terrorism Campaigns

Abigail Watson

Introduction

The failure of two costly military interventions in Iraq and Afghanistan to establish expected levels of stability has led some commentators to announce the death of nation-building as a tool for achieving stability.[1] Placing comparable numbers of Western 'boots on the ground', except in the case of a direct threat to state survival, does not seem likely – at least for another generation when national budgets may have recovered and the memory of Iraq and Afghanistan is more distant.[2] Despite these restrictions, governments remain concerned about potential terrorist threats emanating from places like Iraq, Syria, Libya, Somalia and Nigeria. In this strategic

1 Doug Bandow, 'The Nation-Building Experiment That Failed: Time for U.S. to Leave Afghanistan', *Forbes*, March 1, 2017: https://www.forbes.com/sites/dougbandow/2017/03/01/the-nation-building-experiment-that-failed-time-for-u-s-to-leave-afghanistan/#35f9a24c65b2.

2 Rachael Gribble et al., 'British Public Opinion after a Decade of War: Attitudes to Iraq and Afghanistan', *Politics* 35, no. 2 (June 1, 2015): 128–50; BBC, 'David Cameron: 'Syria Is Not like Iraq', BBC News, July 26, 2016 : http://www.bbc.co.uk/news/uk-politics-23883970; Economist, 'Missing in Action', *The Economist*, March 8, 2014, http://www.economist.com/news/britain/21598654-britain-needs-strategy-make-best-use-its-shrinking-military-capabilities-it-isnt; Richard Norton-Taylor, 'UK Military Operations since Cold War Have Cost £34bn, Says Study', *The Guardian*, April 23, 2014: https://www.theguardian.com/world/2014/apr/23/uk-military-operations-costs.

environment, states like the United Kingdom (UK), the United States (US) and France work with local (often non-state) partners who do the bulk of frontline fighting against terrorist groups such as Boko Haram, al Qaeda and al-Shabaab.

This approach has been a dominant feature of the fight against the so-called Islamic State (IS). In Syria, the US and others supported the Kurdish-dominated Syrian Democratic Forces (SDF). In Libya, the US and France have provided tacit and, potentially, material support to General Khalifa Haftar and the US and the UK supported forces from the Libyan city of Misrata.[3] In doing so, the US, France and the UK each lost fewer than ten soldiers, while the anti-IS coalition 'achieved a singular success' in ousting IS from an area the size of Great Britain.[4] It is, then, perhaps unsurprising that there is a growing narrative around the intervention – seen in the rhetoric of some world leaders and expert commentators – that 'it was a[n operation working 'by, with and through' local partners] that actually worked'.[5] For instance, in March 2018, then-commander of US Central Command (CENTCOM), General Votel, said in evidence to the Senate Armed Services Committee that in these areas the US 'place[d] a heavy reliance on indigenous partner forces…this approach does present its own challenges…[and] is not without risk, as we are seeing unfold in northern Syria today. But it is proving very effective and will pay significant

3 'Missiles Belonging to French Army Found at Pro-Haftar Base in Libya', *France 24*, July 10, 2019, https://www.france24.com/en/20190710-france-missiles-found-pro-haftar-base-libya-un-military-usa-rebels-tripoli.

4 Deborah Haynes, 'British Soldier Captain Dean Sprouting Dies on Iraqi Base', *The Times*, February 3, 2018: https://www.thetimes.co.uk/article/british-soldier-captain-dean-sprouting-dies-on-iraqi-base-c06mvwzzp; 'UK Armed Forces Death: Operational Deaths Post World War II', Ministry of Defence, March 28, 2019: https://assets.publishing.service.gov.uk/government/uploads/system/uploads/attachment_data/file/789825/20190328_UK_Armed_Forces_Operational_deaths_post_World_War_II-O.pdf; Chris Stephen, 'Three French Special Forces Soldiers Die in Libya', *The Guardian*, July 20, 2016: https://www.theguardian.com/world/2016/jul/20/three-french-special-forces-soldiers-die-in-libya-helicopter-crash; Mona Yacoubian, ''By, With, Through' Was the Best Hope for Syria – And Ending 'Endless Wars'' Defense One, October 10, 2019: https://www.defenseone.com/ideas/2019/10/through-was-best-hope-syria-and-ending-endless-wars/160540/; Josie Ensor, 'British Soldier Killed by IED in Syria in First UK Troop Death in Fight against Isil', *The Telegraph*, March 30, 2018: https://www.telegraph.co.uk/news/2018/03/30/british-soldier-killed-ied-syria/.

5 By 'by, with and through' the author refers to operations led by our partners, state or nonstate, with enabling support from international actors (such as the U.S., UK and France); Gayle Tzemach Lemmon, 'The Sense in Syria's Senselessness', War on the Rocks, October 28, 2019: https://warontherocks.com/2019/10/the-sense-in-syrias-senselessness/; 'Analysis: Trump's 'success' in Syria Cedes Region to Russia', Al Jazeera, October 24, 2019: https://www.aljazeera.com/news/2019/10/analysis-trump-success-syria-cedes-region-russia-191024092549326.html.

dividends going forward'.[6] Similarly, in June 2019, US President Donald Trump announced that the US 'ha[s] defeated ISIS in Syria' and made '[t]he world…a much safer place'.[7] It seems, then, that some see these operations as an example of how working with local partners can deliver national objectives with minimal costs in terms of blood and treasure.

This chapter challenges this narrative and argues that, while the recent campaigns against IS had some successes, they also provide a cautionary tale against a 'by, with and through' approach – especially when it comes to working with non-state groups. To explore this argument, this chapter first looks at why working with local non-state groups can be especially problematic, arguing that, by definition, they often threaten state security sectors. Then, the chapter turns to the anti-IS campaign in Syria and in Libya, where the US and its allies worked directly with non-state groups in their efforts to push IS back. In both countries, there were already signs in March 2020 that the way these decisions were developed and managed has had a deeply destabilising impact on the region. In Syria, working with the SDF has exacerbated internal divides, especially in predominantly Arab areas, and increased tension between the US and its NATO ally, Turkey. In Libya, working with Misratan forces has created problems for the United Nations (UN) -backed Government of National Accord (GNA); while supporting General Haftar has posed a direct threat to the GNA's legitimacy and authority and exacerbated internal divisions. These cases show that, while operational successes can be achieved by arming, training and supporting non-state groups, these groups often have diverging long-term aims other than the state supporting them for the country in which they are fighting. Thus, unless arming and training them is part of a larger political strategy which aims to mitigate some of these risks, it can exacerbate political divides and prolong conflict.

---

6 Joseph Votel, 'Testimony Before the Senate Armed Services Committee Transcript', Senate Armed Services Committee, March 12, 2018: https://www.centcom.mil/Portals/6/Documents/Transcripts/13Mar18_Votel_SASC_transcript.pdf.

7 Donald J. Trump, 'We Have Defeated ISIS in Syria, My Only Reason for Being There during the Trump Presidency.', Tweet, *@realDonaldTrump*, June 29 2018: https://twitter.com/realDonaldTrump/status/1075397797929775105; Donald Trump, 'Statement from the President on the Death of Abu Bakr Al-Baghdadi', The White House, October 27, 2019: https://www.whitehouse.gov/briefings-statements/statement-president-death-abu-bakr-al-baghdadi/.

The Dangers of Arming Non-State Actors

Numerous policy documents (such as the UK's 2011 Building Stability Overseas Strategy or the US's 2017 National Security Strategy), public statements and interviews undertaken by the Oxford Research Group (ORG) with key decision makers indicate that there is an enduring assumption that training local and regional forces can allow states like the US and the UK to build local and regional capacity, so that foreign states can address potential threats more autonomously in the future, with few risks of political backlash or to the lives of Western personnel.[8] ORG has spent the last six years examining this assumption and arguing that far from being risk free, these engagements present a number of unique and poorly understood risks and challenges.[9] The risks of working 'by, with, and through' local forces are particularly problematic when it comes to working with non-state groups. According to Hofmann and Schneckener, non-state actors are organisations:

(i) willing and capable to use violence for pursuing their objectives
(ii) not integrated into formalized state institutions such as regular armies, presidential guards, police, or special forces.
(iii) possess a certain degree of autonomy with regard to politics, military operations, resources, and infrastructure.[10]

In many ways, therefore, these groups present threats to a state's 'monopoly of the legitimate use of physical force within a given territory'.[11]

---

8 The White House, 'A New National Security Strategy for a New Era', December 18, 2017, https://www.whitehouse.gov/articles/new-national-security-strategy-new-era/; DFID, FCO, and MOD, 'Building Stability Overseas Strategy' https://assets.publishing.service.gov.uk/government/uploads/system/uploads/attachment_data/file/67475/Building-stability-overseas-strategy.pdf.

9 Emily Knowles and Abigail Watson, 'No Such Thing as a Quick Fix: The Aspiration-Capabilities Gap in British Remote Warfare', *Oxford Research Group*, July 30, 2018: https://www.oxfordresearchgroup.org.uk/no-such-thing-as-a-quick-fix-the-aspiration-capabilities-gap-in-british-remote-warfare; Emily Knowles and Abigail Watson, 'Remote Warfare: Lessons Learned from Contemporary Theatres', *Oxford Research Group*, June 27, 2018: https://www.oxfordresearchgroup.org.uk/remote-warfare-lessons-learned-from-contemporary-theatres; Abigail Watson and Megan Karlshoej-Pedersen, 'Fusion Doctrine in Five Steps: Lessons Learned from Remote Warfare in Africa', *Oxford Research Group*, November 6, 2019: https://www.oxfordresearchgroup.org.uk/fusion-doctrine-in-five-steps-lessons-learned-from-remote-warfare-in-africa.

10 Claudia Hofmann and Ulrich Schneckener, 'Engaging Non-State Armed Actors in State and Peace-Building: Options and Strategies', *International Committee of the Red Cross*, October 16, 2015: https://www.icrc.org/en/doc/assets/files/review/2011/irrc-883-schneckener.pdf

11 Max Weber, *The Vocation Lectures*, ed. David Owen and Tracy B. Strong, trans. Rodney Livingstone, 2/14/04 edition (Indianapolis: Hackett Publishing Company, Inc., 2004).

Additionally, they may also use force against the state they are in or a neighbouring state.

The arming and training of non-state groups by intervening states is not new. In fact, Cunningham, Gleditsch, and Salehyan found that between 1945 and 2011 external actors provided explicit or alleged support to 48 per cent of 443 rebel groups engaged in armed conflict.[12] However, contemporary support of non-state actors does present new challenges; as Hart notes, 'non-state armed groups…are becoming increasingly powerful in vast areas that have been left ungoverned due to failing states'.[13] This is especially true in the aftermath of the so-called Arab Spring where non-state armed groups proliferated as national security sectors fragmented and turned violent against their own population. In this context, international support to non-state groups became a central component of counter-terrorism operations in the region. As Katz argued, there are clear advantages to such relationships – particularly as these groups have 'existing fighting skills, local area knowledge, and proximity to terrorist-held territory' making them 'a ready-made infantry force for foreign powers to train, advise, and enable to assault terrorist targets and clear enemy-held terrain'.[14] A number of military and civilian leaders and experts have pointed to the absence of such elements – local knowledge and existing fighting skills – as significant causes of the failure of the campaigns in Iraq and Afghanistan.[15] As such, it seems reasonable that these elements have now been prioritised.

However, if these relationships with partners are not developed and managed carefully, they risk exacerbating internal divisions and potentially undermining the long-term prospects for peace and stability. This is for two reasons. First, while the US and its allies may share the short-term objective of defeating a terrorist group, it is likely that non-state groups differ in

---

12 Daniel Byman, 'Five Myths about Sponsor-Proxy Relationships', *Lawfare*, December 16, 2018: https://www.lawfareblog.com/five-myths-about-sponsor-proxy-relationships.

13 Laurel Hart, 'Syria's Non-State Armed Groups: A State or UN Problem?', *UNA-UK*, March 23, 2017: https://www.una.org.uk/magazine/1-2017/syria%E2%80%99s-non-state-armed-groups-state-or-un-problem.

14 Brian Katz, 'Imperfect Proxies: The Pros and Perils of Partnering with Non-State Actors for CT', *Center for Strategic and International Studies*, January 29, 2019: https://www.csis.org/analysis/imperfect-proxies-pros-and-perils-partnering-non-state-actors-ct.

15 Eliot Cohen, Jan Horvath, and John Nagl, 'Principles, Imperatives and Paradoxes of Counterinsurgency', *Military Review*, March-April, 2006: http://www.au.af.mil/au/awc/awcgate/milreview/cohen.pdf; Steve Dobransky, 'Why the U.S. Failed in Iraq', *Middle East Quarterly*, January 1, 2014: https://www.meforum.org/3680/iraq-us-failure; Steven Metz, 'Insurgency and Counterinsurgency in Iraq', *The Washington Quarterly* Vol. 27, no. 1 (December 1, 2003): 25–36.

their long-term objectives for a region. As Katz notes: 'Militia may perceive terrorist groups as one among several enemies amid a multi-faceted civil conflict and shift their attention and arms toward other foes when the terrorist threat ebbs'.[16] In such instances, training and equipment provided by states like the US, France or the UK are likely to be used by non-state groups to pursue other threats, sometimes against the wishes of these supporting states. Second, the perception that non-state groups represent a certain ethnicity or area may undermine their legitimacy with the local communities in areas they win back from terrorist groups. Thus, while such groups may help Western forces militarily defeat terrorist groups, they 'may lack legitimacy in the areas they clear and be ill-suited to serve as a holding and stabilization force'.[17] While these groups may be the most militarily able groups, their inability to provide legitimate, accountable and effective security could undermine 'battlefield gains and create[e] openings for terrorists to re-gain local support'.[18]

In this sense, strengthening such groups may serve military objectives but their empowerment may also present problems for building peace after the fighting has ended. In fact, Hofmann and Schneckener argue: 'Non-state armed actors,...often bear the potential to disturb, undermine, or completely truncate processes of peace- and state-building, leading violence to flare up again'.[19] This is not necessarily an argument against working with such groups but, instead, demonstrates the need to carefully manage these relationships and account for the potential long-term consequences. The next section supports this argument by outlining the impact of strengthening non-state groups in Libya and Syria without this careful management of the long-term consequences. Arming non-state groups led to significant military successes against IS but failure to carefully manage the fallout exacerbated instability in the long term.

The Anti-IS Coalition: The Impact of Working with Non-State Groups

In the summer of 2014, IS made rapid territorial gains across Iraq, Syria and Libya. On 17 October 2014, the US Department of Defense established

16 Katz, 'Imperfect Proxies'.
17 Katz,'Imperfect Proxies'.
18 Katz, 'Imperfect Proxies'.
19 Hofmann and Schneckener, 'Engaging Non-State Armed Actors in State and Peace-Building'.

a Combined Joint Task Force – Operation Inherent Resolve (CJTF-OIR), an international coalition focused on defeating IS 'in designated areas of Iraq and Syria and set[ing] conditions for follow-on operations to increase regional stability'.[20] Key to the coalition's strategy was the strengthening of local groups to do the bulk of frontline fighting. As the CJTF-OIR stated: 'Ultimately, the military victory over ISIS will be accomplished by the indigenous forces, we will accomplish our mission with those indigenous forces, and improved regional stability will be attained through those partners'.[21] In Syria, CJTF-OIR supported the SDF, who are controlled by the Kurdish group, the People's Protection Units (Yekîneyên Parastina Gel, YPG). When IS moved into Libya at the end of 2014, the US and others once again focused on empowering local groups to push it back – working, in particular, with forces from the Libyan city of Misrata and General Haftar's Libyan National Army (LNA).[22] In both countries, the territorial gains of IS have been dramatically reduced; however, at the same time, the arming of non-state groups has exacerbated domestic and international divides and worsened the long-term prospects for peace and stability.

Arming the SDF in Syria

When IS moved to take Kobani, a Syrian Kurdish town on the border with Turkey, air support from the US and its allies to YPG fighters on the ground pushed IS back and the siege of the town was successfully broken. As one commentator noted, this was the moment when 'an operational relationship was born. The midwife was tactical necessity. Larger issues of national security objectives, overall strategy for Syria, and an important bilateral relationship with a NATO partner were made subordinate to the singular focus on attacking ISIS'.[23] This relationship continued throughout the fight against IS and in May 2017 the Trump administration directly

20 Combined Joint Task Force – Operation Inherent Resolve, 'Combined Joint Task Force Operation Inherent Resolve: Fact Sheet': https://www.inherentresolve.mil/Portals/14/Documents/Mission/20170717-%20Updated%20Mission%20Statement%20Fact%20Sheet.pdf?ver=2017-07-17-093803-770.

21 Combined Joint Task Force – Operation Inherent Resolve, 'Combined Joint Task Force Operation Inherent Resolve: Fact Sheet',

22 Alison Pargeter, 'After the Fall: Views from the Ground of International Military Intervention in Post-Gadhafi Libya', *Remote Warfare Programme*, July 31, 2017: https://www.oxfordresearchgroup.org.uk/after-the-fall-views-from-the-ground-of-international-military-intervention-in-post-gadhafi-libya.

23 Stein, 'Partner Operations in Syria: Lessons Learned and the Way Forward', *Atlantic Council*, July 2017: http://www.atlanticcouncil.org/publications/reports/partner-operations-in-syria.

armed the YPG for the impending fight for Raqqa. As a 'lessons learnt' report into partner operations in Syria concluded, 'while the SDF has been a tactical success [it has]...created broader political problems', including worsening internal divides and relations between Turkey and the West.[24]

Locally, the fact that many people see the SDF as run and dominated by the Kurdish YPG has made it hard for it to build legitimacy in Arab-majority areas. Initially, the YPG concentrated on majority-Kurdish areas; however, as IS spread, the US and its allies encouraged the YPG to go beyond its own areas of interest and fight IS in some Arab-majority areas where the group often struggled to gain local legitimacy. Despite re-branding itself the SDF and encouraging more Arab recruits, Haid notes 'many Syrian Arabs saw the SDF's attack on rebel-held areas...as a Kurdish pretext to take advantage of US support and expand their territories in areas where Arabs are a majority'.[25] In some areas the SDF and its political affiliate, the Syrian Democratic Council, set up bureaucratic structures in liberated territory to govern areas after the fighting had stopped. However, Crisis Group found that '[o]utside majority-Kurdish areas...[this] governance model appear[ed] fragile'.[26] They describe efforts 'to achieve Arab buy-in to its project' as 'partial and haphazard' rather than 'meaningful'.[27] Haid goes further, stating that 'reported violations committed by some Kurdish groups against Arab communities, have led to ethnic tensions between local communities'.[28] For example, in 2015 there were reports that Syrian Kurds were forcibly removing people from their homes by 'deliberately demolishing civilian homes, in some cases razing and burning entire villages'.[29]

A similar problem has also been evident within SDF's own military ranks. The SDF made impressive efforts to make their ranks more representative of the local community 'and to guard against the region's history of intercommunal violence'.[30] For instance, it has diversified to

24 Aaron Stein, 'Partner Operations in Syria: Lessons Learned and the Way Forward'.

25 Haid Haid, 'The Deal between Syrian Arabs and Kurds That Could Be the Way Forward in Northern Syria', *Middle East Eye*, May 15, 2017: http://www.middleeasteye.net/columns/deal-between-syrian-arabs-and-kurds-could-be-way-forward-northern-syria-274041731.

26 'Fighting ISIS'.

27 'Fighting ISIS'.

28 Haid Haid, 'The West's Obsession with IS Has Not Helped Syria', *CNN*, April 7, 2017: https://www.cnn.com/2017/03/15/opinions/strategy-syria-mess-haid-haid-opinion/index.html.

29 Roy Gutman, 'Have the Syrian Kurds Committed War Crimes?', *The Nation*, February 7, 2017: https://www.thenation.com/article/have-the-syrian-kurds-committed-war-crimes/.

30 Heras and Thomas, 'Solving the Syrian Rubik's Cube'.

include a 'notable number of Arab recruits'.[31] At first, this was relatively successful. The fight for Kobani has been described as a 'successful example of interethnic cooperation under the leadership of the YPG'.[32] However, despite this progress, the SDF 'remains squarely under YPG command and wholly reliant upon the [Kurdish Workers Party's (Partiya Karkerên Kurdistanê, PKK)] trained Kurdish fighters who form its backbone'.[33] As a result, tensions remain. In April 2017, Crisis Group argued that '[t]housands of Arabs now participate in YPG-led military efforts…But they are doing so on behalf of an organisation whose militantly secular culture clashes with local norms, whose Kurdish identity many view as a threat and which has shown no inclination to share power meaningfully'.[34] Even before President Donald Trump announced an imminent US withdrawal, there had been a number of assassination attempts against SDF officials and their local, mainly Arab, allies.[35]

These tensions have only worsened in recent months, particularly with the US's (still relatively unclear) new policy on Syria which will see it dramatically reduce its forces but stay to protect oil fields. Local populations in formerly SDF held territory are now squeezed between Turkish forces and forces loyal to the Assad regime.[36] Since December 2019, this has led to over one million internally displaced persons.[37] SDF has been seen as, at best, powerless to stop this escalating violence and, at worst, complicit in it. For instance, in Deir Ez-Zor province, near the Iraqi border, residents have been protesting against the SDF. They have been demanding 'better services, political representation and access to the oil-rich region's resources'.[38] These protests began in several villages and

---

31 'Fighting ISIS: The Road to and beyond Raqqa', *Crisis Group*, April 28, 2017: https://www.crisisgroup.org/middle-east-north-africa/eastern-mediterranean/syria/b053-fighting-IS-road-and-beyond-raqqa.

32 Nicholas Heras and Kaleigh Thomas, 'Solving the Syrian Rubik's Cube', *CNAS*, April 29, 2019: https://www.cnas.org/publications/reports/solving-the-syrian-rubiks-cube.

33 'Fighting ISIS'.

34 'Fighting ISIS'.

35 Heras and Thomas, 'Solving the Syrian Rubik's Cube'.

36 'The US Changed Its Mind Regarding a Total Withdrawal from Syria', *Middle East Headlines*, March 30, 2019, https://middleeastheadlines.com/the-us-changed-its-mind-regarding-a-total-withdrawal-from-syria/.

37 Matthew Ayton, 'Amid US Uncertainty in Syria, Kurdish YPG Eyes Bolstering Ties with Russia', *Atlantic Council*, March 23, 2020, https://www.atlanticcouncil.org/blogs/menasource/amid-us-uncertainty-in-syria-kurdish-ypg-eyes-bolstering-ties-with-russia/.

38 'Arab Residents Rally against Kurdish Rule in Syria's Deir Az Zor', *Al Jazeera*, May 8, 2019: https://www.aljazeera.com/news/2019/05/arab-residents-rally-kurdish-rule-syria-deir-az-zor-190508061316059.html.

spread to the east of the Euphrates river where most of the oilfields are located. Protesters have been accusing the SDF of selling oil to the regime to enrich their own forces, without giving any to the local population. One protestor said: 'We are deprived of everything while the Kurds are selling our oil to help the regime and enriching themselves'.[39] The protests turned violent in May 2019, when some protesters disrupted convoys from nearby oilfields heading towards government-held areas.[40] Now, these feelings of injustice may be exacerbated by the perception that the US only protects Kurdish oil fields, and not the people within its territory.[41]

Internationally, Western support to the SDF has created problems for the relationship between the West and Turkey. In 1980, following a military coup in Turkey, many leaders within the PKK – a group leading an armed insurgency against the Turkish state – fled to Syria.[42] Now, many in Turkey believe that the leadership of the Syrian Kurdish Democratic Union Party (Partiya Yekîtiya Demokrat, PYD) – who form the armed wing of the YPG – is affiliated with the PKK.[43] As a result, Turkey considers the decision by the US and others to arm the YPG to be 'unacceptable'.[44]

Initially, the US attempted to work with the Turkish government in spite of this conflict of interest. For instance, on several occasions President Barack Obama asked President Recep Tayyip Erdoğan to control the Turkish border with Syria, through which IS fighters flowed freely, but Turkey did little to stop it. Even after the US began working with the SDF, it attempted to placate Erdoğan. For example, it sent military strategists to Ankara to devise a plan to liberate Raqqa with fighters from the Turkish-backed Syrian opposition. However, when 'it became clear that a joint plan with Turkey would require as many as 20,000 US troops on the ground' it was rejected,

39 'Kurdish-Led Authority: Damascus Stirring Tensions in East Syria', *Middle East Monitor,* May 14, 2019: https://www.middleeastmonitor.com/20190514-kurdish-led-authority-damascus-stirring-tensions-in-east-syria/.

40 'Arab Residents Rally against Kurdish Rule in Syria's Deir Az Zor'.

41 Steve Balestrieri, 'Treading Carefully: How American Troops Are Protecting Kurdish Oil Fields', *SOFREP*, March 28, 2020, https://sofrep.com/news/treading-carefully-how-american-troops-are-protecting-kurdish-oil-fields/.

42 'Syria's Kurds: A Struggle Within a Struggle', International Crisis Group, January 22, 2013: http://www.refworld.org/pdfid/510285552.pdf.

43 'Syria's Kurds'.

44 Corey Dickstein 'US Begins Arming Syrian Kurdish Fighters for Raqqa Assault', *Stars and Stripes,* May 30, 2017: https://www.stripes.com/us-begins-arming-syrian-kurdish-fighters-for-raqqa-assault-1.471208?utm_source=Sailthru&utm_medium=email&utm_campaign=EBB%2005.31.2017&utm_term=Editorial%20-%20Early%20Bird%20Brief#.WTAcp2jyvb2.

initially by Obama and then by Trump.[45] Thus, faced with the threat of IS but lacking the will to deploy more troops, the US and its allies worked more closely with the SDF.[46]

The decision to do so has meant that tensions between Turkey and its NATO allies have worsened as the fight against IS has continued. By July 2017, a Turkish state news agency released the locations of ten US military bases in northern Syria, including listing the numbers of troops in some areas, and revealed the presence of French Special Forces in other areas.[47] At the end of 2017, Erdoğan announced that Turkey had signed a deal to purchase a Russian S-400 surface-to-air missile system. Moreover, NATO officials reported that they had not been briefed on the purchase.[48] To make matters worse, Turkey's state-run news agency, Anadolu, published an infographic 'showcasing the types of US aircraft the S-400 system was capable of shooting down'.[49] In response, the US has threatened to block an F-35s sale to Turkey – which a Kremlin spokesman called 'unacceptable'.[50] In January 2018, Turkey launched an air-ground operation against Kurdish forces in the Afrin district of Syria, in response to US efforts to build a Kurdish led 'border security force' – which would have positioned 'potentially… thousands of Kurdish militia fighters along Turkey's southern border'.[51] A two-month military campaign saw Turkish-backed militias take over Afrin in a major blow to Kurdish forces. Before this offensive, Afrin had remained

45 Brett McGurk, 'Hard Truths in Syria', *Foreign Affairs*, May 1, 2019: https://www.foreignaffairs.com/articles/syria/2019-04-16/hard-truths-syria.
46 McGurk 'Hard Truths in Syria'.
47 Roy Gutman, 'Turkey Leaks Secret Locations of U.S. Troops in Syria', *The Daily Beast*, July 19, 2017: http://www.thedailybeast.com/turkey-leaks-secret-locations-of-us-troops-in-syria; 'US Increases Military Posts Supporting PKK/PYD in Syria', *Anadolu Agency*, July 18, 2017, http://aa.com.tr/en/middle-east/us-increases-military-posts-supporting-pkk-pyd-in-syria/864274.
48 Carlotta Gall and Andrew Higgins, 'Turkey Signs Russian Missile Deal, Pivoting From NATO', *The New York Times*, September 12, 2017: https://www.nytimes.com/2017/09/12/world/europe/turkey-russia-missile-deal.html.
49 Shawn Snow, 'Special Ops Commander: No Link between US-Backed Syrian Fighters, Kurdish Terror Group', *Military Times*, October 31, 2017: https://www.militarytimes.com/flashpoints/2017/10/31/special-ops-commander-no-link-between-us-backed-syrian-fighters-kurdish-terror-group/.
50 'Russia Condemns 'US Ultimatum' to Turkey over S-400 Missile Deal', *Al Jazeera*, May 22, 2019: https://www.aljazeera.com/news/2019/05/russia-condemns-ultimatum-turkey-400-missile-deal-190522144753344.html.
51 Eric Schmitt, 'Turkey's President Assails U.S.-Trained Kurdish Border Force', *The New York Times*, January 15, 2018: https://www.nytimes.com/2018/01/15/us/politics/syria-turkey-kurds-border.html.

relatively untouched by the war; afterwards, the situation deteriorated significantly.[52]

Some of these consequences were predictable outcomes of the US working more closely with the SDF, but the limited and, at times, erratic response from the US indicates that it did not properly prepare for them.[53] This was certainly the conclusion of many commentators when, in December 2018, Trump announced an imminent withdrawal of US troops from Syria, following a telephone call with Erdoğan.[54] For nearly a year his advisers attempted to delay this decision to buy 'time for US diplomats to secure a long-term arrangement that might reasonably satisfy Turkey' while at the same time protecting the SDF.[55] However, in October 2019, Trump showed his unwillingness to wait. Following another call with Erdoğan, Trump once again announced an imminent withdrawal of US forces from the country (although, as already noted, US troops have still not fully withdrawn and appear to have stayed to guard oil fields in eastern Syria).[56] In response to Trump's announcement, Turkey launched an offensive to retake SDF controlled areas.[57]

This has endangered the SDF, a key ally in the fight against IS. In the weeks after the announcement, the SDF lost much of the territory they had reclaimed from IS over the previous five years to the Turkish offensive, with civilian casualties reported on both sides. One commentator noted that this resulted in a 'power vacuum' which has seen 'the Kurds, Syrian government troops loyal to President Assad, their Russian backers, ISIS, and Turkish-backed rebel militias vying for control of the northern Syrian territory'.[58] It was estimated that, within a few weeks, 259 Kurdish fighters

---

52 'Turkish-Backed Rebels Looting in Afrin', *BBC News*, March 19, 2018: http://www.bbc.co.uk/news/world-middle-east-43457214; Carlotta Gall and Anne Barnard, 'Syrian Rebels, Backed by Turkey, Seize Control of Afrin', *The New York Times*, March 19, 2018: https://www.nytimes.com/2018/03/18/world/middleeast/afrin-turkey-syria.html; Alex Rossi, 'Kurds Say Syrian City of Afrin Is Being 'ethnically Cleansed' by Turkish Military', *Sky News*, March 10, 2018: https://news.sky.com/story/kurds-say-syrian-city-of-afrin-is-being-ethnically-cleansed-by-turkish-military-11283062.

53 Heras and Thomas, 'Solving the Syrian Rubik's Cube'.

54 Abigail Watson, 'The Myth of Risk-Free Warfare: Lessons from Syria', Inkstick, June 10, 2019: https://inkstickmedia.com/the-myth-of-risk-free-warfare-lessons-from-syria/.

55 McGurk, 'Hard Truths in Syria'.

56 Abigail Watson and Megan Karlshoej-Pedersen, 'It's Not All Trump's Fault: Syria Shows the Danger of War on the Cheap', Defense One, November 4, 2019: https://www.defenseone.com/ideas/2019/11/its-not-all-trumps-fault-syria-shows-danger-war-cheap/161047/.

57 Reese Erlich, 'Russia Is the Only Winner in Syria', *Foreign Policy*, October 2019: https://foreignpolicy.com/2019/10/30/russia-is-the-only-winner-in-syria/.

58 Balestrieri, 'Treading Carefully: How American Troops Are Protecting Kurdish Oil Fields'.

and 120 Kurdish civilians had been killed in northern Syria, while 176,000 'have been forcibly displaced in a region that is already struggling to offer protection for forced migrants'.[59] Mazloum Abdi, the military leader of the SDF, warned of 'ethnic cleansing of the Kurdish people from Syria'.[60] Stuck between Turkish and Syrian forces, the SDF was forced to align with the Assad regime and their Russian ally.

In fact, one of the biggest benefactors of these tensions has been Russia, which 'has leveraged its influence with Ankara and Damascus to emerge as kingmaker'.[61] In October 2019, Erdoğan and Russian President Vladimir Putin reached an agreement to move the SDF out of northern Syria and, instead, to have Russian and Turkish troops jointly patrol a strip of Syria near the border. President Putin's press-secretary, Dmitry Peskov, warned that Russian troops could pull back and allow Turkish forces to 'maul' the SDF if the group did not cooperate.[62] The Russian army has boasted of its critical role. For instance, a Russian officer told the Tass news agency: 'When the Russian flag appears, combat stops – neither Turks nor Kurds want to harm us, so fighting stops thanks to our work'.[63] Now, alongside the S-400 sale, 'the first part of a new Russia-Turkey natural gas pipeline is complete, and Russia plans to build four nuclear power plants in Turkey'.[64] Marcus notes that 'Russia has translated its long-running military support for the Syrian regime into its newly found status as the essential external player in Syria'; while, in contrast, Erlich argues that 'Trump has managed to destroy US credibility, anger allies, and embolden enemies – all without actually bringing US soldiers home as promised in his 2016 presidential campaign'.[65]

Nor is just US influence at stake; many have warned that the recent flare up of violence may also facilitate the resurgence of IS.[66] Detention centres that hold thousands of IS fighters were run by the SDF so, as the SDF are driven out of their territory, there may be escapes and, with

59 Watson and Karlshoej-Pedersen, 'It's Not All Trump's Fault'.
60 Ben Hubbard, 'As U.S. Leaves Allies in Syria, Kurdish Commander Struggles with Fallout', *The New York Times*, October 20, 2019: https://www.nytimes.com/2019/10/20/world/middleeast/pelosi-syria-turkey-jordan.html.
61 Erlich, 'Russia Is the Only Winner in Syria'.
62 Erlich, 'Russia Is the Only Winner in Syria'.
63 Erlich, 'Russia Is the Only Winner in Syria'.
64 Erlich, 'Russia Is the Only Winner in Syria'.
65 Jonathan Marcus, 'Russia and Turkey on Same Page – but for How Long?', *BBC News*, October 23, 2019: https://www.bbc.com/news/world-middle-east-50152390; Erlich, 'Russia Is the Only Winner in Syria'.
66 Heras and Thomas, 'Solving the Syrian Rubik's Cube'.

them, an influx of IS fighters joining the battlefields engulfing northern Syria.[67] McGurk argues that in such a 'vacuum…extremist groups…would thrive'.[68] In fact, reports now suggest that 'ISIS has become a guerilla force' and is 'still active, especially further to the south'.[69] Not only have the US and its allies, arguably, facilitated this rise, their diminished influence also minimises their ability to rectify it.

This is not necessarily an argument against supporting the SDF but it does indicate that the relationship could have been managed better. The consequences of supporting the group were not unforeseen and politicians from across the political spectrum, former and serving military personnel and numerous civil society groups warned against the potential long-term cost of such activities. This could have been mitigated by ensuring such support was couched within a broader strategy which included diplomacy with Turkey and Arab communities and, more generally, a properly thought through exit strategy. Kurdish researcher, Abdulla Hawez, recently argued 'all this could have been avoided if the US had possessed a coherent strategy to prepare the region politically for its eventual withdrawal'.

Supporting Misratan Brigades and General Haftar in Libya

Similar problems are evident in anti-IS operations in Libya, where Western states like the US, the UK and France publicly supported the UN-backed government but, at the same time, empowered non-state groups. This has undermined the GNA's legitimacy and strengthened direct threats to its authority, to the detriment of peace and stability. The fall of General Muammar Gaddafi, following the 2011 NATO intervention, saw the opening of a power vacuum and, with it, a proliferation of non-state groups. Some estimate that up to 1,600 militia groups existed by 2014.[70] Alongside this, state institutions and security forces which had been deliberately kept weak during Gaddafi's rule were unable to provide stability. Pargeter described how Libya was 'in the grip of a vast array of local forces and powerbrokers whose ultimate loyalty lay with their commander, their

---

67 Watson and Karlshoej-Pedersen, 'It's Not All Trump's Fault'.
68 McGurk, 'Hard Truths in Syria'.
69 Balestrieri, 'Treading Carefully: How American Troops Are Protecting Kurdish Oil Fields'.
70 Vincent Durac, 'The Role of Non-State Actors in Arab Countries after the Arab Uprisings', *IEMed Mediterranean Yearbook*, 2015: https://www.iemed.org/observatori/arees-danalisi/arxius-adjunts/anuari/med.2015/IEMed%20Yearbook%202015_NonStateActorsArabUprisings_VincentDurac.pdf.

town or area, their tribe, or in some instances with a particular ideology'.[71] As tensions between some larger groups escalated in mid-2014 violence broke out around Tripoli and Benghazi.[72]

The UN launched a peace process to try to bring the warring factions together in September 2014. In December 2015, some of these factions signed the UN-backed Libyan Political Agreement, which created the GNA, headed by the Presidency Council led by Fayez al-Sarraj.[73] Although dogged by suspicion of Western interference, the GNA was – at the time – the best hope of a unified government in Libya.[74] However, amidst the instability, IS gained a foothold in the country.

Over the course of 2015 and into early 2016, IS took Sirte and expanded its presence into several surrounding towns and villages. It also carried out attacks in some nearby towns and against energy infrastructure in Libya's Oil Crescent. Initially, some countries looked at ways to address this by working with the GNA; for instance, Italy and the UK considered an 'assistance mission…to train the Libyan armed forces, if requested by' the GNA.[75] To strengthen the GNA, attempts were made to absorb militias into a GNA-led state force, yet 'armed groups either resisted becoming part of the state, or agreed to come under the command of the state in a nominal capacity only'.[76]

Despite the precarious position of the GNA, the UK, the US and France began working directly with some of the non-state actors – such as groups from the city of Misrata and General Hafta – already engaged in the fight against IS, through training, intelligence and air support, circumventing the GNA.[77] Wehrey and Badi note that many groups 'presented themselves to the United States and European powers as partners on counter-terrorism…

71 Pargeter, 'After the Fall'.
72 Pargeter. 'After the Fall'.
73 'The Libyan Political Agreement: Time for a Reset', *Crisis Group,* November 4, 2016: https://www.crisisgroup.org/middle-east-north-africa/north-africa/libya/libyan-political-agreement-time-reset.
74 Alison Pargeter, 'After the Fall: Views from the Ground of International Military Intervention in Post-Gadhafi Libya', *Remote Warfare Programme,* July 31, 2017: https://www.oxfordresearchgroup.org.uk/after-the-fall-views-from-the-ground-of-international-military-intervention-in-post-gadhafi-libya.
75 Louisa Brooke-Holland, 'British Troops to Libya?', Briefing Paper (House of Commons Library, April 22, 2016).
76 'Divided We Stand: Libya's Enduring Conflicts', *International Crisis Group,* September 4, 2012: https://www.crisisgroup.org/middle-east-north-africa/north-africa/libya/divided-we-stand-libya-s-enduring-conflicts.
77 Pargeter, 'After the Fall'.

and even security guarantors of the weak central [GNA]'.[78] The US and others, fearing the spread of IS but unable to work with an official state force and unwilling to deploy their own forces, began working with some of the groups who were nominally aligned with the GNA. One of the most powerful of these was a collection of brigades from the Libyan city of Misrata.[79] These brigades formed during the revolution against Gaddafi and, as a result, already 'had several years of fighting experience in Libya's civil war', making them 'the most effective fighting force'.[80] When IS began to expand in the country Misratan groups 'initially sought to contain rather than degrade' it; however, when IS took over Sirte and looked set to expand towards Misrata, the city's militias 'join[ed] the GNA-backed Bunyan al-Marsous (BAM) coalition to retake Sirte'.[81] Katz notes that Misratan militias 'spearhead[ed] the BAM coalition's campaign coastal advance to Sirte and urban combat operations inside the city'.[82] General Joseph F. Dunford Jnr, then chairman of the Joint Chiefs of Staff, said in January 2016 that the US would take 'decisive military action' against IS in Libya, marking the beginning of British and American special forces conducting reconnaissance missions and working alongside Misratan forces.[83] In May 2016, *The Times* reported that UK special forces had been supporting Misratan forces on the frontline against IS.[84] It cited Mohamed Durat, a commander in Misrata, who explained: 'My unit works just with English…I have met with them personally and they have destroyed two suicide vehicles that were targeting my fighters'.[85] In July that year, the US started to conduct airstrikes, 'intended to help break an impasse between Libyan militias and the Islamic State fighters'.[86] When this was announced,

78 Frederic Wehrey and Emadeddin Badi, 'Libya's Coming Forever War: Why Backing One Militia Against Another Is Not the Solution', *War on the Rocks*, May 15, 2019: https://warontherocks.com/2019/05/libyas-coming-forever-war-why-backing-one-militia-against-another-is-not-the-solution/.
79 Pargeter, 'After the Fall'.
80 Katz, 'Imperfect Proxies'.
81 Katz, 'Imperfect Proxies'.
82 Katz, 'Imperfect Proxies'.
83 Helene Cooper, 'U.S. Conducts Airstrikes Against ISIS in Libya', *The New York Times*, August 1, 2016: https://www.nytimes.com/2016/08/02/us/politics/us-conducts-airstrikes-against-isis-in-libya.html.
84 'British special forces destroyed Islamic State trucks in Libya, say local troops', *The Telegraph*, May 26, 2016: http://www.telegraph.co.uk/news/2016/05/26/british-special-forces-destroyed-islamic-state-trucks-in-libya-s/
85 'British special forces fighting in Libya: Report', *Middle East Eye*, May 26, 2016: http://www.middleeasteye.net/news/british-special-forces-fighting-libya-report-201027144.
86 Cooper, 'U.S. Conducts Airstrikes Against ISIS in Libya'.

military officials said that US special forces would 'not call in the airstrikes... Instead...Libyan ground forces, who were given rudimentary instruction by American Special Operations trainers on the ground, provided basic identification and targeting information for the warplanes'.[87] A few months later, Durat was quoted again saying: 'We co-ordinate with the Americans and British. When they are targeting a site we move back, they strike, and we move forward further. Together we can complete this mission'.[88]

The BAM coalition – supported by US air power – successfully retook Sirte from IS in December 2016; however, working with local groups directly undermined the legitimacy of the GNA and exacerbated internal divisions. The BAM coalition collapsed shortly after retaking Sirte over diverging views about serving the GNA and it failed to effectively pursue IS who 'retrenched and regrouped in the mountainous desert south of the city'.[89] Unlike other groups within the coalition, many Misratan brigades have tended to support the GNA; however, this support cannot be taken for granted. Initially, Misratans were opposed to the GNA, controlling a rival political force. They also hoped to advance their own political aims by engaging in the battle against IS. As Katz notes: '[K]ey Misratan militias viewed the Sirte campaign as an opportunity to become Libya's dominant military and political force, not as a purely [counter-terrorism] mission, and not to extend the writ of the...GNA'.[90] However, the prospect of being the armed force of an internationally recognised organisation prompted some factions within the Misratan forces to back the GNA.[91] This provided them with a more 'respectable' status, and, as importantly for many Misratans, served to undermine Haftar and the Libyan National Army.[92] The GNA have never exercised much control over Misratan groups – even though some are aligned with it. Serraj, the Prime Minister of the GNA, lamented in November 2016: 'They do as they please...Whenever they want to go out

---

87 Cooper, 'U.S. Conducts Airstrikes Against ISIS in Libya',

88 'Dramatic moment ISIS jihadis laid down their arms after British special forces helped Libyan fighters flush them out', *The Daily Mail*, 6 August 2016: https://www.dailymail.co.uk/news/article-3727370/Dramatic-moment-ISIS-jihadis-laid-arms-British-special-forces-helped-Libyan-fighters-flush-out.html

89 Katz, 'Imperfect Proxies'.

90 Katz, 'Imperect Proxies'.

91 Pargeter, 'After the Fall'.

92 Alison Pargater, 'Libya's Downward Spiral', *openDemocracy*, February 8, 2015: http://www.opendemocracy.net/alison-pargeter/libya%E2%80%99s-downward-spiral.

and fight, they don't ask us and we end up firefighting these battles'.[93] In fact, some have claimed that these forces continue to threaten the authority of the GNA. For instance, journalist Jalal Othman argued that '[t]hose who were fighting on the ground made benefit of finishing off IS to empower themselves against the state and outside of its authority'.[94]

By directly supporting Misratan groups rather than going through the GNA, Western forces have shown 'how little [the GNA] was actually able to deliver to the forces on the ground in terms of weapons, money or political support', with 'armed groups from Misrata [doing] most of the heavy-lifting, with some help from foreign forces'.[95] While Misratan forces have been 'among the strongest and most powerful supporters of the GNA', and were quick to mobilise against Haftar's offensive in April 2019 (discussed below), the relationship has remained tense.[96] For instance, clashes in Tripoli in 2017, including 'hostile protests against Misrata and violent attacks against its...citizens', led the Misrata local government to suspend its contacts with some leaders of the GNA.[97] This also speaks to the fact that Misratan brigades lack widespread legitimacy in Libya. For instance, some tribes in Sirte see the Misratan-led victory as less of a liberation and more of a conquest. In fact, Wehrey and Lacher argue that 'it was their grievances against Misratan domination that gave ISIS an opening in the first place'.[98] In many ways, then, exacerbating these internal divides by solely supporting Misratan forces may contribute to a resurgence of IS. Frank Archibald, the former Director of the Central Intelligence Agency's (CIA) National Clandestine Service, said that, in Libya, counter-terrorism successes will be undermined unless the 'hard work of governance' is done from 'the day after military success'. He added that failing to do this

---

93 Sami Zapita, 'Serraj Blames Hafter, Saleh, Elkaber and Ghariani for Libya's Problems', *Libya Herald*, November 2, 2016: https://www.libyaherald.com/2016/11/02/serraj-blames-hafter-salah-elkaber-and-ghariani-for-libyas-problems/.

94 Pargeter, 'After the Fall'.

95 Hayder al-Khoei and Mattia Toaldo, 'After ISIS: How to Win the Peace in Iraq and Libya', *ECFR*, January 4, 2017: http://www.ecfr.eu/publications/summary/after_isis_how_to_win_the_peace_in_iraq_and_libya_7212

96 Paul Taylor, 'France's Double Game in Libya', *POLITICO*, April 17, 2019: https://www.politico.eu/article/frances-double-game-in-libya-nato-un-khalifa-haftar/.

97 Wolfgang Pusztai, 'The Failed Serraj Experiment of Libya', *Atlantic Council*, March 31, 2017: http://www.atlanticcouncil.org/blogs/menasource/the-serraj-experiment-of-libya.

98 Frederic Wehrey and Wolfram Lacher, 'Libya After ISIS', *Carnegie Endowment for International Peace*, February 22, 2017: https://carnegieendowment.org/2017/02/22/libya-after-isis-pub-68096.

'creates an ungoverned space that will significantly enable terrorism, crime and human suffering'.[99]

Alongside empowering groups from Misrata, some Western states also supported General Haftar and his forces, the LNA, although to a lesser degree.[100] Haftar was a military officer in Gaddafi's army who defected in the early 1980s and returned to Libya at the time of the 2011 uprisings. His forces, the LNA, consist of 'a loose array of military officers from the former Libyan army, tribesman and local brigades'.[101] Haftar has presented himself as a powerful force, who could 'purify' the country against extremists. This narrative seems to have convinced many to support him.[102] In January 2016, 'a limited number of French special forces...helped the LNA to push back its opponents (including IS) from many of Benghazi's neighbourhoods'.[103] In July 2016, tapes obtained by *Middle East Eye* seemed to indicate that the UK, France and the US were 'helping...Haftar in his campaign to gain control of eastern Libya from groups he has declared to be 'extremists''.[104] Wehrey also notes that, '[i]n 2017, partly due to military support from the [United Arab Emirates (UAE)] and France, [Haftar] finally won the battle for Benghazi'.[105] In this sense, states like the US and France (and to a lesser extent the UK), are supporting both sides of the conflict. In fact, while France has denied providing support to Haftar, it has said it 'has long supported all established forces engaged in the fight against terrorism, in Libya, in the Tripoli area and in Cyrenaica (the east of the country)'.[106] However, while this support allowed Haftar to defeat some 'violent terrorist groups, including those that had attacked the US diplomatic mission', it also supported a group who was in direct opposition to the GNA. Haftar has even used international support to attack 'factions that supported the state and elections – and by lumping them together, he

---

99 Bennett Seftel, 'ISIS Festers and Grows in Lawless Libya', *The Cipher Brief*, January 26, 2018: https://www.thecipherbrief.com/isis-festers-grows-lawless-libya.

100 Hayder al-Khoei and Mattia Toaldo, 'After ISIS: How to Win the Peace in Iraq and Libya' *ECFR*, January 4, 2017: http://www.ecfr.eu/publications/summary/after_isis_how_to_win_the_peace_in_iraq_and_libya_7212

101 Pargeter, 'After the Fall'.

102 Jon Lee Anderson, 'The Unravelling', *The New Yorker*, February 16, 2015, https://www.newyorker.com/magazine/2015/02/23/unravelling.

103 al-Khoei and Toaldo 'After ISIS: How to Win the Peace in Iraq and Libya'.

104 Karim El-Bar, 'REVEALED: Leaked Tapes Expose Western Support for Renegade Libyan General', Middle East Eye, July 8, 2016, http://www.middleeasteye.net/news/revealed-leaked-tapes-expose-western-support-renegade-libyan-general-185825787.

105 Frederic Wehrey, 'The Conflict in Libya', Carnegie Endowment for International Peace, May 15, 2019: https://carnegieendowment.org/2019/05/15/conflict-in-libya-pub-79160.

106 'Missiles Belonging to French Army Found at Pro-Haftar Base in Libya'.

ended up radicalizing them'.[107] This is in direct opposition to the West's stated objectives.

Arming Haftar is even more problematic than working with Misratan forces because he has continued to reject the legitimacy of the GNA and has long sought to 'seize, rather than share, power'.[108] At the end of 2017, for instance, Haftar stated that the UN-led peace process had expired and, with it, so too had the GNA and the Presidency Council.[109] He also said that his own forces, the LNA, were the 'sole legitimate institution' in Libya.[110] In December 2018, Haftar agreed to accept the results of elections; however, many commentators remained sceptical about whether this would be the case.[111] This scepticism was proved well-founded when, in April 2019, just a few days before the UN was due to hold a conference to establish Libya's 'path to reconciliation and elections', Haftar's forces launched a military campaign to take Tripoli from the GNA.[112] In what they have called *Flood of Dignity*, the LNA deployed thousands of men and heavy artillery from their rear base in eastern Libya into the west of the country with the stated aim of capturing the capital.[113] In July 2019, the World Health Organisation said that 1,048 people, including civilians, had been killed since the start of the offensive, with another 5,558 wounded and over 100,000 displaced.[114] Most shockingly, in July 2019, an airstrike by the LNA killed 53 in a migrant detention centre in Tajura.[115]

Despite this 'brazen grab for power by an unrepentant authoritarian', some countries such as Egypt, Russia and the UAE have militarily supported

107 Frederic Wehrey, 'The Conflict in Libya'.
108 Taylor, 'France's Double Game in Libya'.
109 Wehrey and Lacher, 'Libya After ISIS'.
110 Issandr El Amrani, 'New Risks in Libya as Khalifa Haftar Dismisses UN-Backed Accord', *Crisis Group*, December 21, 2017: https://www.crisisgroup.org/middle-east-north-africa/north-africa/libya/new-risks-libya-khalifa-haftar-dismisses-un-backed-accord.
111 Henry Samuel, 'Rival Libyan Leaders in Paris Agree to Hold Elections in December', *The Telegraph*, May 29, 2018: https://www.telegraph.co.uk/news/2018/05/29/rival-libyan-leaders-paris-agree-hold-elections-december/.
112 Patrick Wintour, 'What Does the Battle for Tripoli Mean for Libya and the Region?', *The Guardian*, April 8, 2019: https://www.theguardian.com/world/2019/apr/08/what-does-the-battle-for-tripoli-mean-for-libya-and-the-region.
113 'Averting a Full-Blown War in Libya', *Crisis Group*, April 10, 2019: https://www.crisisgroup.org/middle-east-north-africa/north-africa/libya/averting-full-blown-war-libya.
114 'WHO: More than 1,000 Killed in Battle for Libya's Tripoli', Al Jazeera, July 9, 2019: https://www.aljazeera.com/news/2019/07/1000-killed-battle-libya-tripoli-190708191029535.html; Frederic Wehrey and Megan Doherty, 'Saving Lives in Libya's Nightmarish War', Carnegie Endowment for International Peace, July 16, 2019: https://carnegieendowment.org/2019/07/16/saving-lives-in-libya-s-nightmarish-war-pub-79512.
115 Wehrey and Doherty, 'Saving Lives in Libya's Nightmarish War'.

Haftar, while France and the US have provided diplomatic support to him.[116] France stopped a European Union statement calling for a ceasefire from mentioning Haftar. The US and Russia also said they could not agree to a UN Security Council resolution condemning Haftar, although the UN did later release a statement condemning the airstrike against the migrant detention centre.[117] In fact, President Trump also went as far as to publicly praise the general's 'significant role in fighting terrorism'.[118] Perhaps as importantly, the GNA has received 'barely lukewarm support from the US and Europe' since the start of the April 2019 offensive.[119] This has strengthened Haftar's hand with both his 'domestic' opponents and rivals in eastern Libya and vis-à-vis the government in Tripoli. Yet, as Wehrey and Badi note, Trump's 'view of Haftar as a...figure who can finally forge a degree of unity' is 'misguided'[120] Salamé, UN envoy for Libya, recently warned states supporting Haftar that he would be unable to bring peace to the country as long as he attempts to rule 'with an iron fist'.[121] Instead, his military offensive has displaced 'thousands and unleash[ed] exclusionary forces such as tribalism and ultraconservative Salafism'.[122] Added to this, his political support base scarcely extends beyond the distinctive eastern region, and even there he remains a problematic figure given his past links to the Gaddafi regime. In April 2019, despite internal divides within groups aligned to the GNA, Haftar's offensive was quickly met with a strong counter-offensive, indicating that there remains a strong opposition to him.[123] Empowering him, then, is likely to lead to further fragmentation and violence.

---

116 Wehrey and Badi, 'Libya's Coming Forever War: Why Backing One Militia Against Another Is Not the Solution'.
117 Gabriela Baczynska and Francesco Guarascio, 'France Blocks EU Call to Stop Haftar's Offensive in Libya', *Reuters*, April 10, 2019: https://uk.reuters.com/article/uk-libya-security-eu-idUKKCN1RM2WN; Julian Borger and Patrick Wintour, "No Coherent Policy': Trump's Scattergun Approach Plunges Libya Deeper into Peril', *The Guardian*, April 30, 2019, sec. World news, https://www.theguardian.com/world/2019/apr/30/trumps-scattergun-approach-haftar-libya-deeper-into-peril-analysis.
118 'Trump Praises Haftar in Apparent Reversal of US Policy on Libya', *Al Jazeera*, April 20, 2019: https://www.aljazeera.com/news/2019/04/trump-calls-haftar-praises-significant-role-terrorism-fight-190419182035115.html.
119 Emily Estelle, 'Russia Is Intervening in Libya. Should We Care?', Critical Threats, November 13, 2019: https://www.criticalthreats.org/analysis/russia-is-intervening-in-libya-should-we-care.
120 Wehrey and Badi, 'Libya's Coming Forever War: Why Backing One Militia Against Another Is Not the Solution'.
121 'UN Libya Envoy Warns against Backing Strongman Haftar', *France 24*, April 29, 2019: https://www.france24.com/en/20190429-un-libya-envoy-warns-against-backing-strongman-haftar-tripoli.
122 Wehrey and Lacher, 'Libya After ISIS'.
123 Taylor, 'France's Double Game in Libya'.

With Western support, Misratan groups and Haftar's LNA greatly diminished IS, however, by working directly with these forces, rather than through the GNA, the UK, the US, France and others have changed the balance of power on the ground and emboldened competitors to the GNA.[124] By directly supporting Misratan forces, international actors arguably undermined efforts to bring aligned militias under the meaningful control of the GNA. By supporting Haftar, the West has empowered an individual who they perceived as a powerful force against extremists, but in reality, has repeatedly challenged the GNA and has, arguably, caused more violent conflict than he has stopped. In doing so, they have undermined any hope the GNA had of uniting non-state groups under a common state security sector. Now, the GNA lacks any 'real coalition of political and armed groups backing it besides the moderate elements in Misrata, which have increasing reservations about the government'.[125] In this tumultuous context, the GNA has also been unable to improve Libya's financial situation which has pushed ordinary Libyans and 'its staunchest supporters' to reconsider their support for it.[126] Despite the fact that IS was pushed back and 'oil production has rebounded to a three-year high, Libya is more polarised and fragmented than ever'.[127] This is likely to benefit IS, who have increased activity since the start of Haftar's April offensive and will continue to do so as long as groups 'previously deployed to contain [it] are...focused on fighting one another'.[128]

Conclusion

In an era of increased scepticism over the utility of sending large numbers of Western troops abroad, there appear to be potential benefits in working by, with and through local partners. Doing so could provide a way to address threats abroad without costly deployments of a country's own forces. In this sense, '[p]roviding military assistance to a domestic opposition group is a means, however imperfect, of combating foreign threats without the human, material and political costs of committing large numbers of troops

124 Pargeter, 'After the Fall'.
125 al-Khoei and Toaldo, 'After ISIS'.
126 Wehrey and Lacher, 'Libya After ISIS'.
127 Wehrey and Lacher, 'Libya After ISIS'.
128 Wehrey, 'The Conflict in Libya'.

to combat'.[129] However, as recent campaigns against IS reveal, while there are some benefits to such an approach, it is also important to understand the distinct challenges it presents. Not only can such activities undermine broader political efforts to build lasting peace in the region, but they could even create greater instability and violence in the future. As '[f]or most of the United States' allies in the Middle East,...their gaze was fixed on the wars after the war against the Islamic State'.[130] Thus, as these groups begin to refocus their gaze to their long-term goals, the legacies of Western choices made in the name of countering IS are likely to loom large in the region for many years to come.

This is especially true where Western forces trained and equipped non-state groups. In many recent campaigns, non-state armed groups have proved the only able force on the ground, where states' own forces have been weak, fragmented or another violent actor. Yet, while this support may have proved the 'least bad option' it has not been unproblematic. Training and equipping non-state armed groups rather than working with, or building up, state institutions can weaken the prospects of states providing their own security more autonomously in the future and embolden potential spoilers to peace-making efforts. Moreover, such local groups tend to represent (or be perceived to represent) the interests of a select area or ethnicity and, as such, rarely enjoy the support of a whole country. Their empowerment can, therefore, further fragment divided societies. This is not necessarily to say that states should not work with these groups, especially in areas where there are not many other options, it is to say that these relationships have potentially more risks and need to be managed more carefully and as part of a longer term strategy.

The problems of failing to do this were evident in the anti-IS campaign in Syria and Libya. In both countries, working with non-state groups led to the emboldening of these groups vis-à-vis others vying for power in the country, including the state security sectors themselves. In Syria, working with the SDF pushed back IS and established enduring governance structures in Kurdish majority areas, but it was not seen as legitimate by Arab communities. Moreover, the perceived links between the SDF and the PKK have meant that support to the group remains unacceptable to the Turkish

129 Patricia Sullivan, 'Is It Ever a Good Idea to Arm Violent Nonstate Actors?', *The Conversation*, July 11, 2017: http://theconversation.com/is-it-ever-a-good-idea-to-arm-violent-nonstate-actors-77652.

130 Robert Malley, 'What Comes After IS?', *Foreign Policy*, July 10, 2017: https://foreignpolicy.com/2017/07/10/what-comes-after-IS-islamic-state-mosul-iraq-syria/.

government. In Libya, by supporting Misratan groups and General Haftar, rather than working through the GNA, Western forces greatly diminished the limited legitimacy the GNA had among local groups and has emboldened potential spoilers, especially Haftar who has long rejected the legitimacy of the GNA. This has potentially exacerbated internal divisions and worsened prospects for peace. None of these consequences were unforeseen, but the urgency of defeating IS led many policymakers to 'subordinate' long-term considerations 'to the singular focus on attacking ISIS'.[131]

In addressing this, the US and its allies must recognise that supporting non-state groups to address a short-term threat like IS is a 'profoundly political [undertaking], with potentially large impacts on power distribution and political arrangements in recipient countries'.[132] As such, this support cannot be treated as a technical or tactical endeavour. Instead, support must be provided as part of a broader political strategy which, among other things, considers the long-term political risks and designs exit strategies which mitigate against them. This is perhaps even truer when aiding non-state armed groups, where greater political and diplomatic effort is required to eventually bring them into state institutions once the fighting has stopped. Unfortunately, during the anti-IS campaign, it seems that the immediate task of defeating IS often consumed strategic discussions, and changes in administration often made (especially US) policies erratic. As a result, efforts to develop long-term exit strategies (which mitigated against the predictable and expected consequences) were poorly thought through or poorly delivered. These experiences show that only by acknowledging and addressing the dangers of working with non-state groups can countries like the US, France and the UK ensure that engagements do not lead to more instability and conflict in the future.

References

Al Jazeera. 'Trump Praises Haftar in Apparent Reversal of US Policy on Libya,' April 20, 2019. https://www.aljazeera.com/news/2019/04/trump-calls-haftar-praises-significant-role-terrorism-fight-190419182035115.html.

131 Stein, 'Partner Operations in Syria: Lessons Learned and the Way Forward', *Atlantic Council*, July 2017: http://www.atlanticcouncil.org/publications/reports/partner-operations-in-syria.
132 Vanda Felbab-Brown, 'Fleeting Convergence: Building Partner Capacity and Militias', *Brookings*, December 6, 2016: https://www.brookings.edu/blog/order-from-chaos/2016/12/06/fleeting-convergence-building-partner-capacity-and-militias/.

Al Jazeera. 'Arab Residents Rally against Kurdish Rule in Syria's Deir Az Zor,' May 8, 2019. https://www.aljazeera.com/news/2019/05/arab-residents-rally-kurdish-rule-syria-deir-az-zor-190508061316059.html.

Al Jazeera. 'Russia Condemns 'US Ultimatum' to Turkey over S-400 Missile Deal,' May 22, 2019. https://www.aljazeera.com/news/2019/05/russia-condemns-ultimatum-turkey-400-missile-deal-190522144753344.html.

Al Jazeera. 'Analysis: Trump's 'success' in Syria Cedes Region to Russia,' October 24, 2019. https://www.aljazeera.com/news/2019/10/analysis-trump-success-syria-cedes-region-russia-191024092549326.html.

Al Jazeera. 'WHO: More than 1,000 Killed in Battle for Libya's Tripoli,' July 9, 2019. https://www.aljazeera.com/news/2019/07/1000-killed-battle-libya-tripoli-190708191029535.html.

Anadolu Agency. 'US Increases Military Posts Supporting PKK/PYD in Syria'. Accessed July 20, 2017. http://aa.com.tr/en/middle-east/us-increases-military-posts-supporting-pkk-pyd-in-syria/864274.

Anderson, L. 'The Unravelling,' *The New Yorker*, February 16, 2015. https://www.newyorker.com/magazine/2015/02/23/unravelling.

Baczynska, G. and F. Guarascio. 'France Blocks EU Call to Stop Haftar's Offensive in Libya'. Reuters, April 10, 2019. https://uk.reuters.com/article/uk-libya-security-eu-idUKKCN1RM2WN.

Balestrieri, S. 'Treading Carefully: How American Troops Are Protecting Kurdish Oil Fields,' *SOFREP*, March 28, 2020. https://sofrep.com/news/treading-carefully-how-american-troops-are-protecting-kurdish-oil-fields/.

Bandow, D. 'The Nation-Building Experiment That Failed: Time for U.S. to Leave Afghanistan'. *Forbes*, March 1, 2017. https://www.forbes.com/sites/dougbandow/2017/03/01/the-nation-building-experiment-that-failed-time-for-u-s-to-leave-afghanistan/#35f9a24c65b2.

BBC. 'David Cameron: 'Syria Is Not like Iraq''.' Accessed July 26, 2016. http://www.bbc.co.uk/news/uk-politics-23883970.

BBC 'Turkish-Backed Rebels Looting in Afrin'. March 19, 2018. http://www.bbc.co.uk/news/world-middle-east-43457214.

Borger, J. and P. Wintour. '"No Coherent Policy": Trump's Scattergun Approach Plunges Libya Deeper into Peril'. *The Guardian*, April 30, 2019. https://www.theguardian.com/world/2019/apr/30/trumps-scattergun-approach-haftar-libya-deeper-into-peril-analysis.

Brooke-Holland, L. 'British Troops to Libya?' Briefing Paper. House of Commons Library, April 22, 2016.

Byman, D. 'Five Myths about Sponsor-Proxy Relationships'. *Lawfare*, December 16, 2018. https://www.lawfareblog.com/five-myths-about-sponsor-proxy-relationships.

CNN. 'Missile Dispute Threatens Cyprus Truce,' January 10, 1997. http://edition.cnn.com/WORLD/9701/10/cyprus/index.html?iref=newssearch.

Cohen, E., J. Horvath, and J. Nagl. 'Principles, Imperatives and Paradoxes of Counterinsurgency'. *Military Review*, March-April (2006). http://www.au.af.mil/au/awc/awcgate/milreview/cohen.pdf.

Combined Joint Task Force – Operation Inherent Resolve. 'Combined Joint Task Force Operation Inherent Resolve: Fact Sheet'. Accessed October 28, 2019. https://www.inherentresolve.mil/Portals/14/Documents/Mission/20170717-%20Updated%20Mission%20Statement%20Fact%20Sheet.pdf?ver=2017-07-17-093803-770.

Cooper, H. 'U.S. Conducts Airstrikes Against ISIS in Libya,' *The New York Times*, August 1, 2016 https://www.nytimes.com/2016/08/02/us/politics/us-conducts-airstrikes-against-isis-in-libya.html.

Crisis Group. 'Divided We Stand: Libya's Enduring Conflicts,', September 4, 2012. https://www.crisisgroup.org/middle-east-north-africa/north-africa/libya/divided-we-stand-libya-s-enduring-conflicts

Crisis Group. 'The Libyan Political Agreement: Time for a Reset,' November 4, 2016. https://www.crisisgroup.org/middle-east-north-africa/north-africa/libya/libyan-political-agreement-time-reset.

Crisis Group. 'Fighting ISIS: The Road to and beyond Raqqa,' April 28, 2017. https://www.crisisgroup.org/middle-east-north-africa/eastern-mediterranean/syria/b053-fighting-isis-road-and-beyond-raqqa.

Crisis Group. 'Averting a Full-Blown War in Libya,' April 10, 2019. https://www.crisisgroup.org/middle-east-north-africa/north-africa/libya/averting-full-blown-war-libya.

Davidson, K. '10 Ways America's Situation in the Middle East Will Get Worse'. *DefenseOne*, October 21, 2019. https://www.defenseone.com/ideas/2019/10/10-ways-americas-situation-middle-east-will-get-worse/160752/.

Daily Mail. 'Dramatic moment ISIS jihadis laid down their arms after British special forces helped Libyan fighters flush them out', 6 August 2016. https://www.dailymail.co.uk/news/article-3727370/Dramatic-moment-ISIS-jihadis-laid-arms-British-special-forces-helped-Libyan-fighters-flush-out.html

Dickstein, C. 'US Begins Arming Syrian Kurdish Fighters for Raqqa Assault,' *Stars and Stripes*, May 30, 2017, https://www.stripes.com/us-begins-arming-syrian-kurdish-fighters-for-raqqa-assault-1.471208.

Dobbins, J., M. A. Poole, A. Long, and B. Runkle. 'Post-9/11 Nation-Building: Afghanistan and Iraq'. In After the War, 85–134. Nation-Building from FDR to George W. Bush. RAND Corporation, 2008.

Dobransky, S. 'Why the U.S. Failed in Iraq'. *Middle East Quarterly*, January 1, 2014. https://www.meforum.org/3680/iraq-us-failure.

Durac, V. 'The Role of Non-State Actors in Arab Countries after the Arab Uprisings'. *IEMed Mediterranean Yearbook*, 2015. https://www.iemed.org/observatori/arees-danalisi/arxius-adjunts/anuari/med.2015/IEMed%20Yearbook%202015_NonStateActorsArabUprisings_VincentDurac.pdf.

Economist. 'Missing in Action'. March 8, 2014. http://www.economist.com/news/britain/21598654-britain-needs-strategy-make-best-use-its-shrinking-military-capabilities-it-isnt.

El Amrani, I. 'New Risks in Libya as Khalifa Haftar Dismisses UN-Backed Accord'. Crisis Group, December 21, 2017. https://www.crisisgroup.org/

middle-east-north-africa/north-africa/libya/new-risks-libya-khalifa-haftar-dismisses-un-backed-accord.

El-Bar, K. 'REVEALED: Leaked Tapes Expose Western Support for Renegade Libyan General'. *Middle East Eye*, July 8, 2016. http://www.middleeasteye.net/news/revealed-leaked-tapes-expose-western-support-renegade-libyan-general-185825787.

Ensor, J. 'British Soldier Killed by IED in Syria in First UK Troop Death in Fight against Isil'. *The Telegraph*, March 30, 2018. https://www.telegraph.co.uk/news/2018/03/30/british-soldier-killed-ied-syria/.

Erlich, R. 'Russia Is the Only Winner in Syria'. *Foreign Policy*, October 2019. https://foreignpolicy.com/2019/10/30/russia-is-the-only-winner-in-syria/.

Estelle, E. 'Russia Is Intervening in Libya. Should We Care?,' *Critical Threats*, November 13, 2019. https://www.criticalthreats.org/analysis/russia-is-intervening-in-libya-should-we-care.

Felbab-Brown, V. 'Fleeting Convergence: Building Partner Capacity and Militias'. Brookings, December 6, 2016. https://www.brookings.edu/blog/order-from-chaos/2016/12/06/fleeting-convergence-building-partner-capacity-and-militias/.

France 24. 'UN Libya Envoy Warns against Backing Strongman Haftar,' April 29, 2019. https://www.france24.com/en/20190429-un-libya-envoy-warns-against-backing-strongman-haftar-tripoli.

France 24. 'Missiles Belonging to French Army Found at Pro-Haftar Base in Libya,' July 10, 2019. https://www.france24.com/en/20190710-france-missiles-found-pro-haftar-base-libya-un-military-usa-rebels-tripoli.

Gall, C. and A. Barnard. 'Syrian Rebels, Backed by Turkey, Seize Control of Afrin'. *The New York Times*, March 19, 2018, sec. World. https://www.nytimes.com/2018/03/18/world/middleeast/afrin-turkey-syria.html.

Gall, C. and A. Higgins. 'Turkey Signs Russian Missile Deal, Pivoting From NATO'. *The New York Times*, September 12, 2017, sec. Europe. https://www.nytimes.com/2017/09/12/world/europe/turkey-russia-missile-deal.html.

Gribble, R., S. Wessley, S. Klein, D. Alexander, C. Dandeker, and N. Fear. 'British Public Opinion after a Decade of War: Attitudes to Iraq and Afghanistan'. Political Studies Association, 2014. https://doi.org/10.1111/1467-9256.12073.

Gutman, R. 'Have the Syrian Kurds Committed War Crimes?' *The Nation*, February 7, 2017. https://www.thenation.com/article/have-the-syrian-kurds-committed-war-crimes/.

Gutman, R. 'Turkey Leaks Secret Locations of U.S. Troops in Syria'. *The Daily Beast*, July 19, 2017, sec. World. http://www.thedailybeast.com/turkey-leaks-secret-locations-of-us-troops-in-syria.

Haid, H. 'The West's Obsession with ISIS Has Not Helped Syria'. *CNN*, April 7, 2017. https://www.cnn.com/2017/03/15/opinions/strategy-syria-mess-haid-haid-opinion/index.html.

Haid, H. 'The Deal between Syrian Arabs and Kurds That Could Be the Way Forward in Northern Syria'. *Middle East Eye*, May 15, 2017. http://www.middleeasteye.

net/columns/deal-between-syrian-arabs-and-kurds-could-be-way-forward-northern-syria-274041731.

Hart, L. 'Syria's Non-State Armed Groups: A State or UN Problem?' *UNA-UK,* March 23, 2017. https://www.una.org.uk/magazine/1-2017/syria%E2%80%99s-non-state-armed-groups-state-or-un-problem.

Haynes, D. 'British Soldier Captain Dean Sprouting Dies on Iraqi Base'. *The Times,* February 3, 2018, https://www.thetimes.co.uk/article/british-soldier-captain-dean-sprouting-dies-on-iraqi-base-c06mvwzzp.

Heras, N. and K. Thomas. 'Solving the Syrian Rubik's Cube'. *CNAS,* April 29, 2019. https://www.cnas.org/publications/reports/solving-the-syrian-rubiks-cube.

Hofmann, C. and U. Schneckener. 'Engaging Non-State Armed Actors in State and Peace-Building: Options and Strategies'. Review article. International Committee of the Red Cross, October 16, 2015. https://www.icrc.org/en/doc/assets/files/review/2011/irrc-883-schneckener.pdf.

Hubbard, B. 'As U.S. Leaves Allies in Syria, Kurdish Commander Struggles with Fallout'. *The New York Times,* October 20, 2019, https://www.nytimes.com/2019/10/20/world/middleeast/pelosi-syria-turkey-jordan.html.

International Crisis Group. 'Syria's Kurds: A Struggle Within a Struggle'. January 22, 2013. http://www.refworld.org/pdfid/510285552.pdf.

Katz, B. 'Imperfect Proxies: The Pros and Perils of Partnering with Non-State Actors for CT'. *CSIS Briefs.* Center for Stratgic and International Studies, January 29, 2019. https://www.csis.org/analysis/imperfect-proxies-pros-and-perils-partnering-non-state-actors-ct.

Khoei, H. and M. Toaldo. 'After ISIS: How to Win the Peace in Iraq and Libya'. *ECFR,* January 4, 2017. http://www.ecfr.eu/publications/summary/after_isis_how_to_win_the_peace_in_iraq_and_libya_7212.

Knowles, E. and A. Watson. 'No Such Thing as a Quick Fix: The Aspiration-Capabilities Gap in British Remote Warfare'. Oxford Research Group, July 30, 2018. https://www.oxfordresearchgroup.org.uk/no-such-thing-as-a-quick-fix-the-aspiration-capabilities-gap-in-british-remote-warfare.

Knowles, E. and A. Watson. 'Remote Warfare: Lessons Learned From Contemporary Campaigns'. Oxford Research Group, June 27, 2018. https://www.oxfordresearchgroup.org.uk/remote-warfare-lessons-learned-from-contemporary-theatres.

Knowles, E. and A. Watson. 'Lawful But Awful? Legal and Political Challenges of Remote Warfare'. Oxford Research Group, May 31, 2018. https://www.oxfordresearchgroup.org.uk/awful-but-lawful-legal-and-political-challenges-of-remote-warfare-and-working-with-partners.

Knowles, E. and J. Matisek. 'Western Security Force Assistance in Weak States'. *The RUSI Journal* Vol. 164, no. 3 (April 16, 2019): 10–21.

Lemmon, G. T. 'The Sense in Syria's Senselessness'. *War on the Rocks.* Accessed October 30, 2019. https://warontherocks.com/2019/10/the-sense-in-syrias-senselessness/.

Middle East Eye. 'British special forces fighting in Libya: Report', May 26, 2016. http://www.middleeasteye.net/news/british-special-forces-fighting-libya-report-201027144.

Middle East Monitor. 'Kurdish-Led Authority: Damascus Stirring Tensions in East Syria,' May 14, 2019. https://www.middleeastmonitor.com/20190514-kurdish-led-authority-damascus-stirring-tensions-in-east-syria/.

Middle East Headlines. 'The US Changed Its Mind Regarding a Total Withdrawal from Syria,' March 30, 2019. https://middleeastheadlines.com/the-us-changed-its-mind-regarding-a-total-withdrawal-from-syria/.

Ministry of Defence. 'UK Armed Forces Death: Operational Deaths Post World War II', March 28, 2019. https://assets.publishing.service.gov.uk/government/uploads/system/uploads/attachment_data/file/789825/20190328_UK_Armed_Forces_Operational_deaths_post_World_War_II-O.pdf.

Malley, R. 'What Comes After ISIS?' *Foreign Policy* (blog), July 10, 2017. https://foreignpolicy.com/2017/07/10/what-comes-after-isis-islamic-state-mosul-iraq-syria/.

Marcus, J. 'Russia and Turkey on Same Page – but for How Long?' BBC, October 23, 2019, sec. Middle East. https://www.bbc.com/news/world-middle-east-50152390.

McGurk, B. 'Hard Truths in Syria,' *Foreign Affairs*, May 1, 2019. https://www.foreignaffairs.com/articles/syria/2019-04-16/hard-truths-syria.

Metz, S. 'Insurgency and Counterinsurgency in Iraq'. *The Washington Quarterly* Vol. 27, no. 1 (December 1, 2003): 25–36.

Norton-Taylor, R. 'UK Military Operations since Cold War Have Cost £34bn, Says Study'. *The Guardian*, April 23, 2014, sec. Politics. https://www.theguardian.com/world/2014/apr/23/uk-military-operations-costs.

Pargeter, A. 'Libya's Downward Spiral'. *openDemocracy*, February 8, 2015. http://www.opendemocracy.net/alison-pargeter/libya%E2%80%99s-downward-spiral.

Pargeter, A. 'After the Fall: Views from the Ground of International Military Intervention in Post-Gadhafi Libya'. *Remote Warfare Programme* (blog), July 31, 2017. https://www.oxfordresearchgroup.org.uk/after-the-fall-views-from-the-ground-of-international-military-intervention-in-post-gadhafi-libya.

Pusztai, W. 'The Failed Serraj Experiment of Libya'. Atlantic Council, March 31, 2017. http://www.atlanticcouncil.org/blogs/menasource/the-serraj-experiment-of-libya.

Rossi, A. 'Kurds Say Syrian City of Afrin Is Being 'ethnically Cleansed' by Turkish Military'. Accessed May 15, 2018. https://news.sky.com/story/kurds-say-syrian-city-of-afrin-is-being-ethnically-cleansed-by-turkish-military-11283062.

Samuel, H. 'Rival Libyan Leaders in Paris Agree to Hold Elections in December'. *The Telegraph*, May 29, 2018. https://www.telegraph.co.uk/news/2018/05/29/rival-libyan-leaders-paris-agree-hold-elections-december/.

Schmitt, E. 'Turkey's President Assails U.S.-Trained Kurdish Border Force'. *The New York Times*, January 15, 2018, sec. Politics. https://www.nytimes.com/2018/01/15/us/politics/syria-turkey-kurds-border.html.

Seftel, B. 'ISIS Festers and Grows in Lawless Libya'. *The Cipher Brief* (blog), January 26, 2018. https://www.thecipherbrief.com/isis-festers-grows-lawless-libya.

Snow, S. 'Special Ops Commander: No Link between US-Backed Syrian Fighters, Kurdish Terror Group'. *Military Times,* October 31, 2017. https://www.militarytimes.com/flashpoints/2017/10/31/special-ops-commander-no-link-between-us-backed-syrian-fighters-kurdish-terror-group/.

Stein, A. 'Partner Operations in Syria: Lessons Learned and the Way Forward'. Atlantic Council, July 2017. http://www.atlanticcouncil.org/publications/reports/partner-operations-in-syria.

Stephen, C. 'Three French Special Forces Soldiers Die in Libya'. *The Guardian,* July 20, 2016. https://www.theguardian.com/world/2016/jul/20/three-french-special-forces-soldiers-die-in-libya-helicopter-crash.

Sullivan, P. 'Is It Ever a Good Idea to Arm Violent Nonstate Actors?' *The Conversation,* July 11, 2017. http://theconversation.com/is-it-ever-a-good-idea-to-arm-violent-nonstate-actors-77652.

Taylor, P. 'France's Double Game in Libya'. *POLITICO,* April 17, 2019. https://www.politico.eu/article/frances-double-game-in-libya-nato-un-khalifa-haftar/.

Telegraph. 'British special forces destroyed Islamic State trucks in Libya, say local troops', May 26, 2016. http://www.telegraph.co.uk/news/2016/05/26/british-special-forces-destroyed-islamic-state-trucks-in-libya-s/

Trump, D. J. 'We Have Defeated ISIS in Syria, My Only Reason for Being There during the Trump Presidency'. Tweet. @realDonaldTrump, Tweet, 2018T06:29. https://twitter.com/realDonaldTrump/status/1075397797929775105.

Trump, D. J. 'Statement from the President on the Death of Abu Bakr Al-Baghdadi'. The White House, October 27, 2019. https://www.whitehouse.gov/briefings-statements/statement-president-death-abu-bakr-al-baghdadi/.

Votel, J. 'Testimony Before the Senate Armed Services Committee Transcript', Senate Armed Services Committee, March 12, 2018: https://www.centcom.mil/Portals/6/Documents/Transcripts/13Mar18_Votel_SASC_transcript.pdf.

Watson, A. 'The Myth of Risk-Free Warfare: Lessons from Syria,' *Inkstick,* June 10, 2019, https://inkstickmedia.com/the-myth-of-risk-free-warfare-lessons-from-syria/.

Watson, A. and M. Karlshoej-Pedersen, 'It's Not All Trump's Fault: Syria Shows the Danger of War on the Cheap,' *Defense One,* November 4, 2019. https://www.defenseone.com/ideas/2019/11/its-not-all-trumps-fault-syria-shows-danger-war-cheap/161047/.

Weber, M. *The Vocation Lectures.* Edited by David Owen and Tracy B. Strong. Translated by Rodney Livingstone. 2/14/04 edition. Indianapolis: Hackett Publishing Company, Inc., 2004.

Wehrey, F. and W. Lacher. 'Libya After ISIS'. Carnegie Endowment for International Peace, February 22, 2017. https://carnegieendowment.org/2017/02/22/libya-after-isis-pub-68096.

Wehrey, F. 'The Conflict in Libya,' Carnegie Endowment for International Peace, May 15, 2019 https://carnegieendowment.org/2019/05/15/conflict-in-libya-pub-79160.

Wehrey, F. and E. Badi. 'Libya's Coming Forever War: Why Backing One Militia Against Another Is Not the Solution'. *War on the Rocks*, May 15, 2019. https://warontherocks.com/2019/05/libyas-coming-forever-war-why-backing-one-militia-against-another-is-not-the-solution/.

Wehrey, F. and M. Doherty, 'Saving Lives in Libya's Nightmarish War,' Carnegie Endowment for International Peace, July 16, 2019. https://carnegieendowment.org/2019/07/16/saving-lives-in-libya-s-nightmarish-war-pub-79512.

Wintour, P. 'What Does the Battle for Tripoli Mean for Libya and the Region?' *The Guardian*, April 8, 2019, sec. World news. https://www.theguardian.com/world/2019/apr/08/what-does-the-battle-for-tripoli-mean-for-libya-and-the-region.

Yacoubian, M. "By, With, Through' Was the Best Hope for Syria – And Ending 'Endless Wars" *Defense One*, October 10, 2019. https://www.defenseone.com/ideas/2019/10/through-was-best-hope-syria-and-ending-endless-wars/160540/.

Zapita, S. 'Serraj Blames Hafter, Saleh, Elkaber and Ghariani for Libya's Problems'. *Libya Herald*, November 2, 2016. https://www.libyaherald.com/2016/11/02/serraj-blames-hafter-salah-elkaber-and-ghariani-for-libyas-problems/.

8

THE EVOLVING CHARACTER OF VIOLENT NON-STATE ACTORS IN MODERN CONFLICT

Perspectives from the Stabilisation Unit, Army Legal Services and the Land Warfare Centre

Julien Bastrup-Birk, Grant Davies and Matthew Walker

Introduction

This chapter is the outcome of a panel on practical responses to the challenges posed by Violent Non-State Actors (VNSA) at the 'Violent Non-State Actors in Modern Conflict' symposium held at the Royal Military Academy Sandhurst. The panel was aimed at providing a realm of exchange between three practitioners with different institutional perspectives (the Stabilisation Unit, the Land Warfare Centre and the Army Legal Services) engaged in dealing with VNSAs. To facilitate the discussion and to provide a forum for developing synergies, sharing best practices and to identify lessons learned, the participants were asked to respond to two set questions:

1. What are the key practical challenges posed by Violent Non-State Actors in modern conflict?
2. What current and future trends do you see developing in relation to violent non-state actors in modern conflict?

Julien Bastrup-Birk (Stabilisation Unit)

What are the key practical challenges posed by violent non-state actors in modern conflict?

VNSAs – including armed movements, terrorist organisations, militant groups and insurgencies – continue to pose a particularly thorny challenge for policy makers and strategic thinkers around the world. The phenomenon, clearly, is not new. Irregular Spanish and Portuguese *Guerrillas*, for example, helped to repel Napoleon's armies during the Peninsular War of 1808-14 in what Carl von Clausewitz described as 'people's war'. 'Citizen soldier' commandos later conducted long-range hit-and-run attacks on British formations during the Boer Wars. In 1914, the Belgrade-based Black Hand organisation effectively triggered World War I, when one of its Serbian nationalist operatives, Gavrilo Princip, assassinated Archduke Franz Ferdinand. World War 2 demonstrated the role that groups of partisans, resistance fighters and guerrillas could play in defeating stronger conventional opponents. In turn, the prevalence of VNSA groups spread rapidly after 1945 within the context of independence, anti-colonial, Cold War, and then post-Cold War conflict to such an extent that most of today's conflicts routinely feature a wide cast list of armed movements and insurgencies fuelled by a broad range of grievances and political ambitions.

Whilst summarising the many policy, strategic and tactical challenges posed by VNSAs in itself constitutes a difficult undertaking, the analysis of different conflicts across regions does highlight the following overarching themes.

A rapidly evolving threat

The armed groups and organisations encountered in today's conflicts tend to react and morph continuously to the evolving political and security context within which they operate.[1] This forms part of a long-standing historical trend in which actors are rapidly forced to adapt in order to survive, particularly when facing stronger opponents such as government forces. This may involve rebranding, restructuring and adopting new (including terrorist) tactics. The example of the Palestinian Liberation Organisation (PLO), although historic, remains relevant to the study of

1 Sebastian. von Einsiedel et al., *Civil War Trends and the Changing Nature of Armed Conflict*, Occasional Paper 10, United Nations University Centre for Policy Research (2017), 3-7.

modern VNSAs. Over thirty years, the PLO saw its headquarters move from Damascus to Amman, and then to Beirut and Tunis, popularised high-profile airliner hijackings and fought a guerrilla war against Israel in Lebanon.[2] Al Qaeda, which grew out of the Afghan Mujahideen resistance to the Soviet Union, experienced a similar evolution, developing a far-reaching financial network that used charities as fronts, establishing cells in countries such as Somalia, Sudan and Yemen and providing training and advice to affiliates around the world.[3] Al Qaeda splinter groups formed the core of subsequent terrorist groups, including al Qaeda in Iraq (AQI), from which the Islamic State in Iraq and Syria (ISIS) later emerged. Armed groups may also suddenly appear as a result of – or aided by – sudden shocks in the international system. Following the collapse of the Gaddafi regime in Libya, weapons looted from the country's stockpiles were transported to Mali by smugglers working for al Qaeda in the Islamic Maghreb (AQIM), where they were used by a coalition of Tuareg rebels and Islamist groups during the country's 2012 crisis.[4]

'Competing' with alternative political structures

The strategies, manifestos and narratives of VNSAs often include a rejection of the status quo, political elite or the presence and/or influence of external powers. Indeed, self-determination and rejectionist narratives provide the justification for contemporary groups such as the Afghan Taleban, Boko Haram and Chechen separatists to name only a few. It comes as no surprise, therefore, that groups regularly attempt to establish alternative governance structures aimed at replacing the state or filling political vacuums. In Somalia, Al Shabaab developed a system of Islamic governance in which controlled regions were administered by local governors – *walis* – under the authority of the organisation's executive *Shura* (governing council).[5] In a variation on this theme, ISIS imposed established *wilayat* ('provinces' or 'states'), each administered by a dedicated ruler, including in geographically

2 See for example *The PLO Offices in Tunis and Amman: Possible Roles, Links and Conduct in Fatah Operation*, Central Intelligence Agency (CIA) Report (Released in April 2014), pp. 2-4; and Kenneth Katzman, *The PLO and its Factions*, Congressional Research Service Report for Congress, The Library of Congress (2002).

3 See *9/11 Commission Report: The Official Report of the 9/11 Commission and Related Publications*, Washington (2004), 171;

4 Zineb Abdessadok, *Libya Today: From Arab Spring to Failed State*, Al Jazeera, May 30, 2017. See also Frederic Wehry, Insecurity and Governance Challenges in Southern Libya, Carnegie Endowment for International Peace Report (2017).

5 Stig Jarle Hansen, *Al-Shabaab in Somalia: The History and Ideology of a Militant Islamist Group, 2005-2012*. (London: Hurst & Company, 2013), 83–90.

dislocated *wilayas* as far away as Libya, Algeria and Afghanistan.[6] Such governance models give rise to a situation where the government may find itself having to 'compete' with the structures and services of non-state actors in contexts in which formal institutions are plagued by bureaucracy, inefficiency, corruption, and where key segments of the population are excluded from the prevailing political deal. These dynamics are doubly problematic: even if the informal structures of non-state actors can be disassembled through security operations, the perceived illegitimacy of the state may inevitably breed renewed attempts at challenging its authority as well as feed competing legitimacy narratives. This risk is particularly acute when non-state actors benefit from freedom of movement, or where they draw on the support of key constituencies as well as economic resources.

Local knowledge and strategic patience

In contrast to the above, a recurring challenge for intervening forces, including those from Western states, is precisely that they are often unable and ill-equipped to operate 'amongst the people'.[7] Non-state actor movements are usually willing to accept higher attrition rates than conventional actors and to 'wait out' their opponents, gradually co-opting formal institutions and expanding their local support base in anticipation of the right moment to consolidate their gains. Mao Zedong, in his early writings on guerrilla warfare, summarised the approach that would be adopted by countless generations of insurgencies, emphasising the importance of securing base areas and conducting indirect attacks as a prelude to more conventional forms of warfare.[8] These principles featured heavily in the thinking of North Vietnamese General, Nguyen Giap,' during the Vietnam war and, later, formed a cornerstone of the Afghan Taleban's overall strategy. Clearly, such strategic patience is most effective in contexts where the government is dependent on external backing for survival and where domestic support for interventions within intervening nations is wavering.

---

6 Jessica Stern and J.M Berger, *ISIS: The State of Terror*. (London: William Collins, 2015), 51. See also Katherine Bauer (ed.), *Beyond Syria and Iraq: Examining Islamic State Provinces*, The Washington Institute for Near East Policy, Washington DC (2016), 15–40.

7 See Rupert Smith, *The Utility of Force: The Art of War in the Modern World*. (New York: Alfred A. Knopf, 2005).

8 See Michael Y. M. Kau and John K. Leung (eds.), *The Writings of Mao Zedong: 1949–1976: Volume I, September 1949 – December 1955*. (New York: M.E. Sharpe Inc., 1986), 632.

Ungoverned spaces and border areas

An additional challenge borne out of recent conflicts is the extent to which armed groups exploit porous borders in order to establish strategic depth and evade capture or defeat. Such an approach is particularly effective in environments where groups' opponents are politically limited to geographical boundaries. The value of cross-border operations and supply lines, which was well understood by 'traditional' insurgents such as the Vietcong,[9] can now be seen within the context of Pakistan's tribal areas or indeed, Boko Haram's incursions into the Lake Chad basin region. Borderlands, particularly those featuring difficult terrain such as mountainous or jungle areas, may also provide safe spaces for training camps and 'out of area' headquarters from which to plan campaigns, launch attacks, fundraise and recruit. The examples of Colombia, Pakistan, Syria and the Sahel also demonstrate the extent to which such areas are also exploited by violent criminal actors. Indeed, criminal groups typically benefit from some of the same conditions as insurgencies, namely freedom of movement (for the purposes of smuggling and commodity production), access to trading hubs (many of which are historically located near borders) and weak or absent formal governance structures. Such groups may not necessarily seek to overthrow the state but do nonetheless have an interest in ensuring that it remains weak or co-opted.[10] In this respect their aims may overlap with that of armed political groups and they may embark in logistical (such as protection in return for funding) or even local political agreements.

What current and future trends do you see developing in relation to violent non-state actors in modern conflict?

Recent and current conflicts, moreover, point to some key evolutions in the character and modus operandi of VNSAs around the world. A few of the most prevailing trends are outlined below.

Migrating tactics and ideas

The study of contemporary VNSAs demonstrates the extent to which armed groups draw on and benefit from ideas, expertise and tactics from

9 Chinese weapons and supplies were transported along the Ho Chi Minh trail, a network of paths crossing through the jungles of Laos and Cambodia, to Vietcong fighters in Vietnam.

10 See for example Annette Idler, *Borderland Battles: Violence, Crime and Governance at the Edges of Colombia's War.* (Oxford, Oxford University Press, 2019).

around the globe. Such a phenomenon, admittedly, is not new. For example, doctrine and training manuals provided to non-state actor groups by the Allies during World War II (many of which were translated into multiple languages) quickly spread to anti-colonial and liberation insurgencies in the post-War period. Similarly, the Mujahideen-era Maktab al-Khidamat ('MAK' or 'Office of Services'), essentially the forerunner to al Qaeda, drew on 1950's CIA 'black book' training guides.[11] However, the speed and spread in the migration of both ideas and tactics has accelerated significantly in recent years, whether with respect to political narratives and grievances, as witnessed within the context of the Arab Spring, or the effectiveness of Improvised Explosive Devices (IEDs) and sensationalist suicide tactics in the modern age of mass and social media coverage.

The above example of the training manuals also hints at a wider trend in which non-state actors have adopted methods and strategies previously developed by states in contexts such as Afghanistan, Iraq, Somalia and Mali. Possibly realising a lack of political appetite on the part of intervening forces to incur casualties, groups in these countries have, in effect, succeeded in applying traditional counter-insurgency principles, such as by 'boxing in' troops and aid workers within 'green zones' and bases in order to isolate them from the local population (in effect a reversal of David Galula's Algeria-era emphasis on separating insurgents from the population).[12]

Technology as a force multiplier

The above trend introduces an additional, interrelated theme: the extent to which technology has provided non-state actors with methods, tools and capabilities enjoyed traditionally only by state actors. The most obvious example is the internet, which offers groups a means of conducting virtual reconnaissance of targets as well as planning operations, establishing funding networks, developing recruiting material and establishing an international brand. Readily available smartphones, radios, video cameras, four-wheel-drive vehicles and motorbikes have all become standard capabilities for armed movements around the world, providing them with both mobility and effective command and control systems. In an additional evolution, low-budget, amphibious maritime and air capabilities have also

11 Andrew McGregor, Jihad and the Rifle Alone: 'Abdullah' Azzam and the Islamist Revolution, *Journal of Conflict Studies* Vol. 23, no. 2 (2006), 107.

12 See for example Gregory Fremont-Barnes (ed.), *A History of Counterinsurgency: From South Africa to Algeria, 1900-1954.* (Santa Barbara, Praeger Security International, 2015), 148–239.

been harnessed at different times by violent groups. For example, the team of ten Lashkar-e-Tayyiba (LeT) terrorists that conducted the 2008 Mumbai attacks sailed to India aboard a hijacked Indian-flagged fishing trawler before disembarking in Mumbai aboard inflatable dinghies.[13] Ten years later, ISIS used commercially available drones for the purposes of aerial surveillance, which were later weaponised.[14]

Violent non-state actors as proxies

An enduring pattern in the evolution of VNSA groups is the extent to which these continue to be used as proxies by state actors in the international system in order to further their foreign policy interests. This trend, which formed a key feature of the Cold War, is beginning to re-emerge after a period of US hegemony within the international system. Examples include Iran's enduring sponsorship of Hezbollah and support for Shia militias in Syria and Iraq, the use of Northern Alliance against the Taliban in Afghanistan or, slightly more unconventionally, the services rendered to Russia by the Luhansk branch of the Russian Night Wolves motorcycle gang (which included manning checkpoints in annexed Crimean territory).[15] Such a trend reflects Hoffman's (2007) notion of 'hybrid' conflicts in which 'a full range of different modes of warfare including conventional capabilities, irregular tactics and formations, terrorist acts including indiscriminate violence and coercion, and criminal disorder'[16] are harnessed by state actors when it is in their strategic interests to do so or when they do not wish to be seen as active participants in a war. In a wider pattern, seen in South Sudan, Libya, Sierra Leone and elsewhere, armed groups, and militias specifically, may also be employed either as a substitute for – or 'contracted' capability of – the state (rather than external powers), particularly in contested areas.

As this brief account has demonstrated, VNSAs, through a combination of historic legacies and contemporary developments pose an enduring challenge to those trying to counter them. Such actors may not, at first glance, possess the strength or resources available to most state actors, but they do have their own advantages: agility, access to local political constituencies, knowledge of the ground and, increasingly,

13 Angel Rabasa *et al.*, *The Lessons of Mumbai*, Occasional Paper, RAND Corporation (2009), 3–8.
14 Charles Clover and Emily Feng, 'ISIS use of hobby drones as weapons tests Chinese makers', *The Financial Times*, December 11, 2017.
15 Jack Losh, 'Putin's Angels: the bikers battling for Russia in Ukraine', *The Guardian*, January 29, 2016.
16 Douglas C. Lovelace Jr., *Hybrid Warfare and the Gray Zone Threat*, Terrorism: Commentary on Security Documents Vol. 141 (Oxford: Oxford University Press, 2016), 153.

technology. Moreover, interventions over the past two decades appear to point to an uncomfortable conclusion: that there are no clear, 'one size fits all' responses to these actors. Whilst planning implications are beyond the scope of this chapter, a growing body of evidence, including stabilisation doctrine highlights the need to understand the political economy of conflict, and how non-state actors interact with and exploit constituencies excluded from local settlements.[17] Indeed, the factors outlined in this chapter hint at the inherent difficulties involved in defeating groups through military means alone. Instead, violent elements are more effectively isolated when the constituencies whom they depend upon for support – and the leaders whom they look to for direction – are brought into the political fold and convinced that there aims are best met through means other than an armed campaign.

Lt Col. Grant Davies (Army Legal Services)

What are the key practical challenges posed by violent non-state actors in modern conflict?

As this book has discussed in detail, recent (but by no means all) conflicts such as those in Syria, Libya, Iraq and Afghanistan have involved irregular and 'privatised' forms of violent actor. Such actors have been categorized as warlords, some as 'mercenaries', others as terrorists, 'mafia' type groups, private military and security companies. In such confused environments, the search for clarity regarding their status is often elusive. Arguably, the extreme complexity of twenty-first century conflicts amplifies the need to clearly define such actors now more than ever.

This short contribution will examine current trends and issues regarding VNSAs from a legal perspective, leaving the political, moral and ethical issues to others, although it is readily recognised that such issues often and inevitably overlap. The conceptual challenges to categorising these actors and the practical implications for shaping conflict is complex, even within the narrow framework of international law. Nonetheless, such critical assessments are vital if states are to retain moral authority and legal certainty, key elements of campaign legitimacy.

---

17 *The UK Government's approach to Stabilisation: A guide for policy makers and practitioners*, UK Stabilisation Unit, March 2019.

International humanitarian law (IHL) or the Law of Armed Conflict (LOAC) relies on, as it its principle frame of reference, the recognition of an armed conflict – international armed conflict (IAC) and non-international armed conflict (NIAC).[18] Anything out with or below that threshold is subject to national or domestic law, including human rights law.[19] To the right of the 'bright red line' of armed conflict a degree of certainty exists based on a body of law and rules that have developed into a framework under which armed forces and states can operate, particularly in international conflict. It is undeniable, however, that there exists a 'grey zone" to the left of this line: it is not always clear where riots, tensions and internal disturbances[20] end and a NIAC (or even an IAC) begins. It is this 'grey zone' which state and NSAs seek to exploit as any uncertainty in the mind of their opponent will usually be to their advantage. States facing western liberal democracies under a rules-based international system (under the United Nations Charter) exploit this delta with the weapon of uncertainty.

Who are 'non-state actors'?

The most widely accepted framework of defining who is entitled to take part in hostilities is within the Geneva conventions, notably Geneva Convention 3 (GC III) which determines who is entitled to PW status. It

18 See Common Articles 2 and 3 of Geneva Convention for the Amelioration of the Condition of the Wounded and Sick in Armed Forces in the Field (First Geneva Convention) (1949), 75 UNTS 31, Geneva Convention for the Amelioration of the Wounded, Sick, and Shipwrecked Members of Armed Forces at Sea (Second Geneva Convention) (1949), 75 UNTS 85, Geneva Convention Relative to the Treatment of Prisoners of War (Third Geneva Convention) (1949), 75 UNTS 135, Geneva Convention Relative to the Protection of Civilian Persons in Time of War (Fourth Geneva Convention) (1949), 75 UNTS. 287.

19 In armed conflict, NIAC or IAC Human Rights is not entirely displaced by the Laws of armed Conflict. ICJ, Legal Consequences of the Construction of a Wall in the Occupied Palestinian Territory, Advisory Opinion, 9 July 2004, ICJ Reports 2004 para.106; 'some rights may be exclusively matters of international humanitarian law; others may be exclusively matters of human rights law; yet others may be matters of both these branches of international law'. See also Human Rights Committee, General Comment 29, States of Emergency (Article 4), UN Doc. CCPR/C/21/Rev.1/Add.11 (2001) para. 3; Concluding Observations of the Committee on Economic, Social and Cultural Rights: Israel; 31/08/2001. E/C.12/1/Add.69.

20 Additional Protocol II (APII), which is applicable only in NIAC states 'This Protocol shall not apply to situations of internal disturbances and tensions, such as riots, isolated and sporadic acts of violence and other acts of a similar nature, as not being armed conflicts'. https://ihl-databases.icrc.org/applic/ihl/ihl.nsf/INTRO/475. For caselaw regarding the definition of NIAC see *Prosecutor v Tadić*, Decision on the Defence Motion for Interlocutory Appeal on Jurisdiction (Appeals Chamber) IT-94-1-AR72 (2 October 1995), *Prosecutor v Tadić* (Judgement) IT-94-1-T, Judgement (Appeals Chamber) (15 July 1999), and *Prosecutor v Haradinaj* (Judgement) IT-04-84-T (3 April 2008).

must be recalled that the GCIII is of course only applicable in IAC and thus may not be relevant *stricto sensu* in NIAC. Of relevance to the discussion, GC III gives PW status (and therefore it can be strongly argued the yoke of state responsibility) to certain persons not forming part of the regularly constituted armed forces. These include (*inter alia*):

> (4) Persons who accompany the armed forces without actually being members thereof, such as civilian members of military aircraft crews, war correspondents, supply contractors, members of labour units or of services responsible for the welfare of the armed forces, provided that they have received authorization from the armed forces which they accompany, who shall provide them for that purpose with an identity card...[21]

Much debate has been held over whether civilian technicians, advisors or even operators of remote operated aerial vehicles are 'members of the armed forces' and therefore come under the authority of the state. A growing trend has also emerged whereby civilian technical personnel advise and assist other nations armed forces, particularly in technical areas such as aircraft arming and maintenance to such an extent arguments have been made that they are in fact forming part of the 'kill chain' and thus arguably taking a part in active hostilities.[22] The extent of this assistance could attract state attribution if not legally, then morally. This risk is increased where such assistance is not via purely private industrial assistance, but where the military forces of a state blur the line between 'advising' and actually 'directing' military action. The much-vexed question of what constitutes the loss of civilian protected status against attack, i.e. Direct Participation of Hostilities (DPH)[23] can be as problematic for the state's own forces as it can be for states targeting others under the same principle

Legal accountability is also regarded by some as a 'rigged game', in favour of Non-State Actors. Within an armed conflict, it is generally accepted that IHL/LOAC applies to both states and NSAs,[24] whereas Human Rights

21 *Supra,* Third Geneva Convention, Art. 4.

22 See for example Aaron Merat. 'The Saudis couldn't do it without us' in *The Guardian,* 18 June 2018, https://www.theguardian.com/world/2019/jun/18/the-saudis-couldnt-do-it-without-us-the-uks-true-role-in-yemens-deadly-war.

23 The International Committee of the Red cross interpretive guidance can be viewed here: https://www.icrc.org/en/doc/assets/files/other/icrc-002-0990.pdf. It is not uncontroversial.

24 Common Article 3 to the 1949 Geneva Conventions (CA3) states that 'each Party to the conflict shall be bound to apply, as a minimum' its provisions, and the 1977 Additional Protocol II (AP II) applies to those NIACs between State forces 'and dissident armed forces or other organized armed groups'.

law (save for the basic protections in Common Article 3) does not.[25] The practical challenges connected with evidence gathering and prosecution are not inconsiderable, not least because the capacity and willingness of the host state is so critical. The mismatch or imbalance regarding the state's obligations towards individuals (including captured insurgents) and the lack of accountability in Human Rights law for NSAs is not lost on governments and militaries alike. The state is accountable under Human rights law, the NSA are not. States can be held accountable for 'NSA' actions but only where there are direct linkages, but this is highly problematic (see below). In such cases, it is questionable whether therefore the NSA is truly an NSA.

Threshold of violence – and the difficulty defining an armed attack by non-state actors

In order to justify an armed response, under the UN system and customary international law, a state must have suffered an 'armed attack'. This may seem straightforward, but what level of violence, or disruption is universally agreed as constituting such an attack that merits an armed response, and by whom? Do only conventional means attract a conventional response or could armed force be used in response to a cyber based attack? How does a state measure proportionality when faced with a non-conventional attack (such as the 9/11 attack on the USA in 2001)? Would a lone actor justify a wholescale state response to a cyber-attack that had kinetic effect? The lack of universal consensus about such triggers can and are exploitable by state and NSAs who may seek to achieve objectives without triggering a costly armed response by, for example, using carefully calculated disruptive measures which fall short of an obvious armed attack. Furthermore, if the waters can be muddied by passing off the attack as one by a 'lone wolf individual' – the rogue state or even non-state actor – could attack with impunity.

Attribution by non-state actors

NSAs rarely operate without some state involvement, less so in the case of 'true' organized crime gangs. The difficulty lies in defining state involvement

25 For a general synopsis of accountability of NSAs in Human Rights law see K. Fortin, *The Accountability of Armed Groups under Human Rights Law* 27–68 (Oxford: Oxford University Press, 2017) and D. Murray, *Human Rights Obligations of Non-State Armed Groups* 120–154 (Oxford: Hart, 2016).

(and therefore attribution) is the extent to which state involvement crosses the line of sympathy, minor logistical support to actual control or direction (the latter arguably attracting state responsibility and retaliatory measures such as armed force). For those states operating under the UN rules-based system, this poses real challenges, particularly in the cyber domain where even geographical identification of a cyber-attack may not necessarily equate to state attribution. There is considerable difficulty in determining precisely who – a government, independent actor or skilled individual – is behind such attacks. Even in a conventional domain, recent events have seen the use of so-called 'little green men',[26] whose national identity was at best confused, or sufficiently confused to delay or cast doubt in any state's mind as to what if any retaliatory measures to take.

Whilst the International Law Commission Draft Articles on State responsibility doubtlessly helped clarify secondary rules concerning attribution to the state of individuals' actions, caselaw on such points is still developing.[27] The bar for attribution is arguably quite high[28] (ie actual control or direction) and the evidence proving the nexus between a state and NSA remains problematic, particularly in the cyber domain. Such obstacles will no doubt give succour to states or transnational NSAs wishing to use such proxies and proto-PMCs to further their interests without the prospect of repercussions.

The paucity and ambiguity of law in NIAC for assisting states regarding non-state actors

One characteristic of NIAC is that persons captured by the state's armed forces are not Prisoners of War (a category that exists only in IAC). These NSAs are internees or detainees, and the usual course is to retain such persons until they are handed to the prosecuting authorities of the state

26 A term used to describe the use of military personnel with no national markings so as to confuse or mask the state behind military action. See Vitaly Shevchenko, 'Little green men' or 'Russian Invaders', BBC news (online) 11 March 2014; https://www.bbc.co.uk/news/world-europe-26532154.

27 See United Nations, International Law Commission, Report on the work of its fifty-third session (23 April–1 June and 2 July–10 August 2001), General Assembly, Official Records, Fifty-fifth Session, Supplement No. 10 (A/56/10), available at http://www.un.org/law/ilc/. There is however a growing body of academic commentary that these rules in fact reflect customary international law, and therefore carry some considerable weight.

28 Various cases have dealt with attribution of state responsibility for non-state actors, the first and most well-known case being the International Court of Justice (ICJ) case of *Military and paramilitary Activities in and against Nicaragua (Nicaragua and the United States of America)* UCJ Rep 1986, 14 ('The Nicaragua case').

in which the conflict arises. The expectation that such detainees would be prosecuted and imprisoned by a functioning legal system is often unrealistic. NIAC frequently occur in fragile states, where support for the rule of law and a functioning judicial, penal and policing system is far from being a top priority. Recent case law (in the UK at least) has shown that the basis for third party states to detain (and retain) insurgents is limited, increasing the likelihood of facing the insurgents again after release, or litigation (and resulting loss of legitimacy) at home when such incarceration is challenged in British courts.[29]

This situation places assisting states in an invidious position and may even prevent them from providing such assistance going forward. This is not necessarily a failing of international law, but a matter of practical reality for which states must find solutions, either by passing of domestic law or entering into treaties which empowers assisting states regarding detention.

Whilst detaining NSAs is difficult, targeting the same is also problematic. Save for the armed forces of the state, there are no 'lawful combatants' in a NIAC, and thus those NSAs who oppose the state do so as effectively criminals. Whilst their legal status is relatively clear, the right to use lethal force against NSAs is mired in controversy and lack of universal agreement as to the basis of such action. The basic rule in IHL is that civilians shall never be directly targeted, but can be when, and for only as long as they are directly participating in hostilities. This seemingly simple rule has, and continues, to provoke debate amongst legal scholars and military practitioners to this day and despite an attempt to provide definitive guidance issued by the ICRC, remains an issue that is far from resolved.

Wither the mercenary, and the rise of the PMC and proto-PMC

The term 'mercenary' is frequently used in the context of NSAs. Although a useful frame of analysis, the accepted international legal definition is drawn very tightly. Of course, a mercenary is an illegal participant in any armed conflict, and there are negative, moral as well as political and legal ramifications for being labelled a mercenary. The definition can be found within Protocol I of the Geneva Conventions. Article 47(2) defines mercenaries as any person who:

29 For a discussion on the difficulties faced by third party intervening states in NIAC see Davies. G, 'Revolving Doors and Human Rights: Detention in Internationalised non-International Armed Conflict' in Brown et al. *War Amongst the People: Critical Assessments*, Howgate Publishing, 2019.

a) is specially recruited locally or abroad in order to fight in an armed conflict;
b) does, in fact, take a direct part in the hostilities;
c) is motivated to take part in the hostilities essentially by the desire for private gain and, in fact, is promised by or on behalf of a Party to the conflict, material compensation substantially in excess of that promised or paid to combatants of similar ranks and functions in the armed forces of that Party;
d) is neither a national of a party to the conflict nor a resident of territory controlled by a Party to the conflict;
e) is not a member of the armed forces of a Party to the conflict; and
f) has not been sent by a State which is not a Party to the conflict on official duty as a member of its armed forces.

The specificity and multiple qualifications required to qualify as a mercenary have produced the quip 'anyone who is charged with being a mercenary should sack their defence attorney'. Certainly, many Private Security Companies would not qualify, mainly on the grounds that they do not 'fight' in the conflict or 'do not take a direct part in hostilities', but rather act as heavily armed security personnel using force only in self-defence. However, it must be stated that such PMCs do not operate in a legal vacuum, they will be subject to the law of the state in which they operate. The real issue is the ability or willingness of the state to prosecute where transgressions are alleged to have occurred.

The fact also remains, in that in a world where many states have reduced their military spending and therefore manpower, the option of 'contracting out' security or capability remains attractive. The perceived ability to 'buy-out' liability by using such PMCs could, however, be illusory. This is a complex area.

Another developing trend is the phenomenon of organisations that appear to be NSAs or PMCs but are in reality acting under the direction of states, or those with very close ties to states.[30] Such 'proto-PMCs' have the obvious attraction of extending state influence without the tie of attribution of legal, moral or ethical responsibilities. Such a benefit is as alluring as it is disturbing.[31]

---

30 See for example the CNN report by Tim Lister, Sebastian Shukla and Clarissa Ward '*Putins Private Army*', regarding the activities of 'Wagner' in the Central African Republic, here: https://edition.cnn.com/interactive/2019/08/africa/putins-private-army-car-intl/.

31 See also '*Russia's hired guns in Africa*' Sergey Sukhankin, European Council on Foreign Relations, https://www.ecfr.eu/article/commentary_russias_hired_guns_in_africa.

How do states counter such actions, and how are such actors held to account? Whilst accountable in the state in which they operate, it is the state itself which may well be paying for such services. The prospect of legal accountability is therefore, slim. Private Security companies are big business, nowhere more so than in the UK[32] and have benefits to states and even organisations such as the UN. The undeniable trend is that PMCs or proto-PMCs are not going away anytime soon, and states must learn to deal with these players in any modern conflict.

What current and future trends do you see developing in relation to violent non-state actors in modern conflict?

Even where an armed conflict is in train, the paucity of concrete rules within NIAC continues to cause frustration. There is little chance of any internationally agreed rules in NIAC, as the same barriers that existed during the drafting of the Geneva Conventions and protocols exist as much today as they did then.[33] The lack of international law rules regarding detention of NSAs in NIAC by assisting states relies on a functioning legal system and the maintenance of the rule of law within the state the conflict occurs. Experience shows that this presumption has little or no foundation where collapsed or collapsing states either lack the will or the capability to police, convict and incarcerate the NSAs. There is neither the will nor a consensus to add to existing rules regarding detention in NIAC – so this issue will continue to pervade.

The use of PMCs shows no sign of abating, and there are very good reasons why states can and do use such outsourcing solutions with relative impunity. 'Good behaviour of the PMCs'[34] relies on the voluntary acceptance of norms of behaviour by the PMCs themselves, and the willingness and

32 See Richard Norton-Taylor, *The Guardian Online*, 3 February 2016, 'Britain is at the centre of global mercenary industry says charity' at https://www.theguardian.com/business/2016/feb/03/britain-g4s-at-centre-of-global-mercenary-industry-says-charity.

33 As was evident during discussions regarding an agreement for rules NIAC detention during the Copenhagen Process, Winkler, T 'The Copenhagen Process on Detainees: A necessity', *Nordic Journal of international Law* 78 (2010).

34 See the 'Montreux Document on Private Military and Security Companies' issued by the ICRC https://www.icrc.org/en/publication/0996-montreux-document-private-military-and-security-companies. According to its sponsor, the ICRC, it is *'an intergovernmental document intended to promote respect for international humanitarian law and human rights law whenever private military and security companies are present in armed conflicts. It is not legally binding as such but, rather, contains a compilation of relevant international legal obligations and good practices. The Montreux Document was finalized by consensus on 17 September 2008 by 17 States, and the number of participating States is growing steadily'*.

ability of host states to prosecute where transgressions occur. Perhaps even more worryingly is the potential re of proto-PMCs, ie clandestinely state funded NSAs which exploit the weaknesses in enforcing 'good behaviour' and the outward lack of state attribution to their activities. Such benefits will not be lost on those states seeking to bolster their power and influence outside their borders.

Attribution of state sponsored NSA actions to the sponsor states remains elusive and perhaps only the development (for example, by caselaw if not hard rules) could stem the tide. The use of proxies is not new and developments such as the Draft Articles of State responsibility[35] are welcome, but cases in this field are complex, and difficult, but not entirely impossible, to enforce.[36]

Finally, whilst this volume rightly identifies the issue of the prevalence of non-state actors as a growing trend in all armed conflicts, states should not fall into the trap of 'fighting the last war'. Focussing too heavily or exclusively on this phenomenon invites the danger of losing corporate memory in fighting large scale conventional conflicts, both physically and conceptually.

Lt Col Matthew Walker (Land Warfare Centre).

What are the key practical challenges posed by violent non-state actors in modern conflict?

The British Army's interventionist operations since the turn of the millennium have sought to provide stability in fragile states affected by the malign influence of VNSAs. Adopting a lessons-learned approach, this contribution examines what aspects of the challenges presented by VNSAs in recent conflict are likely to endure and assesses how the Army is adapting to this. Whether the experience of these conflicts was new or

35 http://legal.un.org/ilc/texts/instruments/english/commentaries/9_6_2001.pdf, in particular Article 16 'Article 16. Aid or assistance in the commission of an internationally wrongful act A State which aids or assists another State in the commission of an internationally wrongful act by the latter is internationally responsible for doing so if: (a) that State does so with knowledge of the circumstances of the internationally wrongful act; and (b) the act would be internationally wrongful if committed by that State.

36 For an example of such complexity and difficulty see Marko Milanovic article online at EJILTalk! '*Russian agents charged with downing of MH17; MH 17 cases in Strasbourg*', at https://www.ejiltalk.org/russian-agents-charged-with-downing-of-mh17-mh17-cases-in-strasbourg/.

unique may be debated; what is important is that how to operate within this environment, initially at least, appeared to be beyond the corporate memory of the conventional forces intervening in these conflicts. This principal observation shows that drawing on appropriate lessons from the past and blending these with a nuanced understanding of the threat is central to the art of conducting effective military operations against VNSAs. The lessons learnt by military forces in targeting the VNSA's narrative, adapting to the VNSA's ever-changing *modus operandi* and in coordinating all levers of government in a campaign to defeat the VNSA are discussed.

Operating through, and within, indoctrinated or aligned members of the wider community, VNSAs' capacity to take ownership of the operational and strategic[37] narratives presents a significant practical challenge. The VNSAs western forces have faced in recent conflicts have principally sought to impose their own beliefs and to achieve their own desired outcomes. Such groups have been motivated by advancing their ideological, governance or economic aspirations. The instability these groups cause may support the objectives of states seeking to challenge the rules-based world order and as a result they may have attracted the fiscal, military or ideological support of these states. The challenge presented by VNSAs acting as state proxies is discussed later.

Targeting VNSAs is extremely challenging. Where permissible in law, conventional military forces may readily find and defeat any militaristic apparatus the VNSA may establish or may deny infrastructure; however, the VNSA does not place the same importance on these traditional objectives. The VNSA places much more emphasis, and a group's survival is more dependent, on cognitively influencing supporters and non-aligned observers through traditional and social media and broader cyber operations. In this respect VNSAs are defending a dogma or ideal. And these ideals may change, develop or evolve as military operations progress.

While a 'positional' or geographic defeat may be achieved against a VNSA, defeat of the dogma or 'idea' they espouse and promulgate invariably cannot; thus, a series of tactical or operational military successes against a group may not translate to its strategic ideological or political

37 I have used these terms as they are defined in British military doctrine (Land Warfare Development Centre; Army Doctrine Publication 'Land Operations' (AC71940)). The operational level is the level at which campaigns and major operations are planned, conducted and sustained to accomplish strategic objectives within theatres or areas of operations. The tactical level is the level at which activities, battles and engagements are planned and executed to accomplish military objectives assigned to tactical formations and units.

defeat. The challenge of permanently defeating the VNSA is exemplified in operations against the so-called Islamic State of Iraq and the Levant (ISIS) – a group that, initially at least, adopted a traditional, geographic-based positional defence of the territory it held using organised, albeit irregular, forces. As operations against ISIS intensified – a development that weakened the organisation significantly – ISIS's response was to fight in a less traditional, more asymmetric manner, exploiting fear amongst its ranks, employing suicide IEDs and conducting violent raids against weaker elements of the attacking force. When geographic withdrawal was no longer possible the group simply refocused its efforts in the cyber domain to cohere support, message its ideology and, thereby, assure its organisational survival. Recent operations thus demonstrate that, when it can, a non-state group will seek to conduct itself with the demeanour of a state actor, but when it is unable to do so, it continues to exploit all the levers of influence available to it, but manipulates them differently, presenting itself as a coherent guerrilla group, cellular insurgent group, or social agitator thereafter. Consequently, where the threat a group presents emerges next, and in what form, remains uncertain.

Furthermore, VNSAs have been remarkably adroit and agile, exploiting readily available consumer technology, most vividly illustrated in modifying drones intended for amateur photography to provide a disposable surveillance capability and means to deliver modified conventional munitions from the air.[38] To understand such amorphous opponents, when not actively engaged in operations, organisations such as the Land Warfare Centre reach out to partners to draw on operational analysis of strategic trends, and to observe operations more broadly in areas of instability to learn about war by watching others. This is supported by the UK's Defence Attaché network and reporting from the plethora of UK-led capacity building deployments.

Leveraging all traditional levers of governmental power to defeat the VNSA has proved problematic. The report of the Chilcott Enquiry into the UK operations in Iraq[39] articulates the limitations of conducting military action without effective, coordinated and complementary mechanisms to promote economic development and accountable governance. In line with the 'whole of government' approach espoused in the government's Fusion

38 US Army Mosul Study Group, 'What the Battle for Mosul teaches the Force', September 2017.

39 'The Report of the Iraq Enquiry: Executive Summary. Report to the Privy Counsellors'. House of Commons, 6 July 2016.

Doctrine[40] military action is increasingly seen to be supporting activity that must be synchronised with other levers of national power. The British Army's conceptual framework has been adjusted to reinforce this. This is evidenced in the publication of the most recent series of field manuals for stability operations.[41]

Within these publications, the British Army has broadened its approach to counter-insurgency and is now additionally considering the contribution it may make to counter-terrorism and criminality, thus recognising VNSAs' adeptness in fusing their activities across these three paradigms. Similarly, refined doctrine for the military contribution to stabilisation provides tactical commanders with guidance on how they may synchronise the delivery of military effect with the wider cross-government effort to achieve stability.[42]

What current and future trends do you see developing in relation to violent non-state actors in modern conflict?

Focus within the British Army is increasingly turning to training for warfighting operations. In doing this, it will be important institutionally not to ignore the enduring threat VNSAs may present and the lessons of recent campaigning against them. While their capacity to emerge in ungoverned space or to harness the dissatisfaction of populations whose human security needs are failed by traditional states is undiminished, how future threats will emerge and where is difficult to predict. Accordingly, this analysis will consider the likely nature of future inter-state conflict more generally and emerging trends in the role VNSAs may take in these conflicts, drawing distinctions between analysis which considers specific state threats and those which more broadly consider 'grey zone' activity.

Contemporary military discussion is energetically considering the potential for conventional force-on-force conflict in the foreseeable future, perhaps not as the *most likely* basis for future expeditionary operations, but certainly as the *most dangerous* scenario. Future state-on-state conflict is

---

40 HM Government, National Security Capability Review, March 2018 which introduces the Fusion Doctrine with the aim to 'to improve our collective approach to national security…so that we use our security, economic and influence capabilities to maximum effect to protect, promote and project our national security, economic and influence goals'.

41 Land Warfare Centre Army Field Manual (AFM), 'Tactics for Stability Operations (TFSO)' (AC72063), 2017; AFM TFSO Part 1: Counter Irregular Activity', 2019; 'AFM TFSO Part 3: The Military Contribution to Humanitarian Assistance and Disaster Relief' (AC72134), 2019; and 'AFM TFSO Part 4: The Military Contribution to Stabilisation' (AC 72135), 2019.

42 Land Warfare Centre Army Field Manual.

likely, however, to take a different form, where unconventional, irregular and proxy forces have a more significant role, to the industrialised clash of uniformed military forces. Recognising the meandering flux in intensity of 'conflict' between states, the British Army now describes a state of 'constant competition'.

The 'grey zone' has become a significant area of study[43] and maximising the competitiveness of conventional military forces in this space is likely to be key to maintaining an advantage in the next decade. Its genesis may be found in the Russian Gerasimov doctrine for exploiting multiple levers of effect what has become popularly known as the 'grey zone' of activity below the threshold of traditional conventional conflict.[44] In implementing this doctrine for what has been termed 'ambiguous warfare',[45] Russia's engagement in the Balkans since 2014 has 'resembled dispatches coming from the trenches of political and economic warfare: Moscow has slashed gas supplies, banned imports of agro-food products, conducted coordinated disinformation campaigns, nurtured nationalist organisations, deployed Cossack paramilitary groups, tested cyber defences and allegedly even tried to overthrow legitimate governments'.[46] This approach clearly shows the breadth of potential roles for the VNSA in inter-state competition.

While thought-provoking, current analysis does not define this zone consistently.[47] Current conceptual military thought relating to the 'grey zone', which includes consideration of some military activity below the NATO Article 5 threshold for warfare and 'multi-domain' operations,[48] generally falls into two areas: non-traditional operations conducted generically in times of 'constant competition' at all levels of intensity (the

---

43 Elizabeth G Troeder: 'A Whole-of-Government Approach to Gray Zone Warfare' (sic) (Strategic Studies Institute and U.S. Army War College Press), May 2019, typifies the debate on the competitiveness of conventional forces and the existential threat of 'grey zone' activity on the national security of western states.

44 Described in the works of Kathleen Hicks, John Schaus, and Michael Matlaga; Linda Robinson et al; Michael Mazarr; and Nathan Freier et al.; see the references in the endnotes to this chapter. The concept of a 'grey zone' and the challenge it presents to western governments is also discussed broadly elsewhere, within security think tanks and in the doctrine branches of NATO and European armies.

45 AE Goldberg, '*Bespredel* and the Conduct of Russian 'Hybrid Operations', *Small Wars Journal* (2016).

46 Stanislav Secrieru, 'Russia in the Western Balkans – Tactical Wins, Strategic Setbacks', EUISS.

47 Discussion at the RUSI conference on 'Securing Competitive Advantage' reflects this. Jack Watling and Nick Reynolds, 'Conference Report Royal United Services Institute for Defence and Security Studies Land Warfare Conference Securing Competitive Advantage' (RUSI), June 2019.

48 TRADOC Pamphlet 525-31, 'The US Army in Multi-Domain Operations 2028', November 2018.

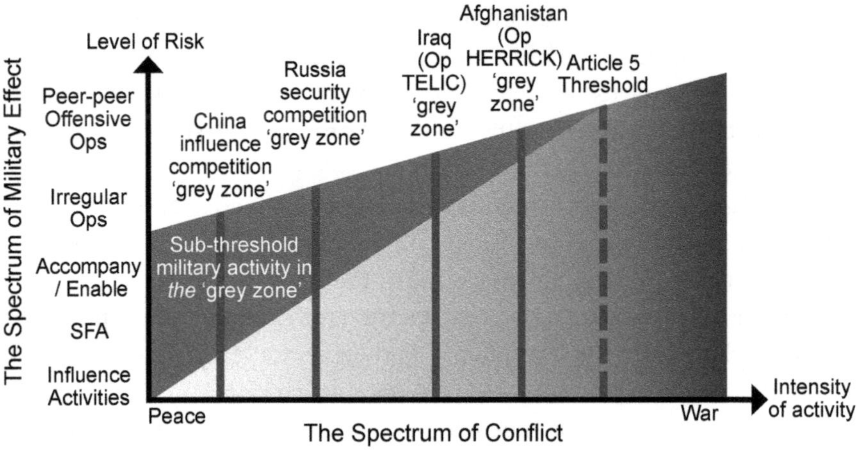

Figure 8.1: ***A*** **'grey zone' or** ***the*** **'grey zone'?**
Author's own analysis and graphic representation. In part drawn from analysis of the 'grey zone' in Army Field Manual: Tactics for Stability Operations, Part 1 Counter Irregular Activity (2018), page 1–2, and internal discussion within the Land Warfare Centre on the nature of the 'grey zone' and its codification in military doctrine. Drawn from Chapters 1, 2 and 3 of the British Field Manual Tactics for Stability Operations Part 1: Counter-irregular Activity.

'grey zone'); or conventional military capacity building or deterrence activity[49] seeking to influence a stated adversary, usually Russia, China, North Korea or Iran (a series of separate 'grey zones') (see Figure 8-1). A common feature to these analyses is the challenge of facing an adversary state who has blended the activities of VNSAs within their wider approach to gain advantage.

Deniable, multi-domain actions short of conventional combat offer leverage in an era of 'constant competition'. Conceptual studies currently define either generic approaches synchronising these levers (*the* 'grey zone') or strategies addressing discrete state threats (*a* series of 'grey zones').

The trend for ambiguous and deniable action by malign states is likely to continue and with it see an increase in the use of VNSA as state proxies, as well as increasingly significant relationships between malign states, organised crime groups, warlords, self-determination groups,

49 Michael Rühle: 'Deterring Hybrid Threats: the need for a more Rational Debate' (NATO Defence College), July 2019 argues that, 'Most hybrid actors, at least state actors, do have a face and an address. Consequently, deterrence needs to be tailored to each specific instance', 3.

militias, and other irregular actors who may contribute to uncertainty and instability within a target area of influence.[50] Understanding how malign state actors may leverage VNSAs to support their objectives by proxy and how this may be addressed by conventional and non-traditional military capabilities is likely to present a considerable challenge to military planners in the future. As states refine how they leverage the greatest impact with the resources available to them, the significance of remote systems, covert special force operations, offensive cyber operations, and overt military capacity building operations is likely to grow exponentially.

Regardless of the conflict or context, a military truism endures: conflict rarely calls for military action at a convenient time or for military capabilities that exactly match those systems held in a force's inventory. Adaption, flexibility and making best use of available resources are key. This will be especially important when engaged in operations against VNSAs who are likely to continue to exploit their ready access to consumer technology to improvise systems to defeat high tech, high spec, and high cost conventional military hardware. The British Army is currently engaged in extensive work to optimise its structure to face these emerging threats, enabling readiness to 'fight tonight'.

The British Army's approach has traditionally been rooted in the physical dimension. In the future, however, niche military capabilities and the coercive impact of the military's hard power are likely to be equally significant in defeating the will or capability of emerging adversaries, whether VNSAs, VNSAs as proxies for state actors, or, conventional forces, to fight. Out of this the conceptual foundation for and restructuring needed to deliver 'information manoeuvre' has arisen. This recognises that the use of information can also achieve a position of advantage, either by enabling physical manoeuvre or by affecting the virtual dimension directly. The concept of information manoeuvre has been determined and is described in emerging doctrine as:

> the harmonisation of land information capabilities across defence's federated capabilities, which are: Cyber and Electromagnetic Activities Intelligence, Surveillance and Reconnaissance, Counterintelligence, Information Activities and Outreach, and Communication and

50 Drawn from Chapters 1, 2 and 3 of the British Army Field Manual Tactics for Stability Operations Part 1: Counter-irregular Activity; the content of these chapters of the field manual are in turn drawn from relevant entries on the (UK) Defence Lessons Information Management System, engagement across the (UK) Field Army and international partners and broad engagement with other government departments and academia.

> Information Systems. By exploiting the synergies between these capabilities in the physical and virtual dimensions, audience cognition is affected in accordance with commander's intent to gain or maintain information advantage, which in turn supports the integration of actions.[51]

In future operations where VNSAs present a significant threat, it is clear that effectiveness will demand an approach which exploits all diplomatic, economic, cultural and military levers of national power and ever greater cross-government input in military planning to articulate the military contribution to a coordinated, 'whole of government' response.

References

9/11 Commission Report: The Official Report of the 9/11 Commission and Related Publications, Washington: Government Publication, 2004.

Abdessadok, Z. *Libya Today: From Arab Spring to Failed State,* Al Jazeera, May 30, 2017.

Bauer, K. (ed.). *Beyond Syria and Iraq: Examining Islamic State Provinces.* Washington DC: The Washington Institute for Near East Policy, 2016.

(British Army) Land Warfare Centre, Army Field Manual (AFM), 'Tactics for Stability Operations (TFSO)' (AC72063), 2017.

(British Army) Land Warfare Centre, AFM TFSO Part 1: Counter Irregular Activity' (AC 72136), 2019.

(British Army) Land Warfare Centre, AFM TFSO Part 4: The Military Contribution to Stabilisation' (AC 72135), 2019.

Central Intelligence Agency (CIA) Report. *The PLO Offices in Tunis and Amman: Possible Roles, Links and Conduct in Fatah Operation,* October 1985, Released April 2011.

Clover, C. and E Feng. 'ISIS use of hobby drones as weapons tests Chinese makers', *The Financial Times,* December 11, 2017.

Davies. G. 'Revolving Doors and Human Rights: detention in Internationalised non-International Armed Conflict' in Brown, *War Amongst the People: Critical Assessments,* Ed. D. Brown et al., Howgate Publishing.

Dennys, C. *Military Intervention, Stabilisation and Peace – the Search for Stability,* New York: Routledge, 2016.

Dobbins, J. *Extending Russia. Competing from Advantageous Ground,* Santa Monica: RAND, 2019.

Flanagan, S. *Deterring Russian Aggression in the Baltic States through Resilience and Resistance,* Santa Monica: RAND, 2019.

51 Land Warfare Centre, Information Manoeuvre Doctrine Note 19/XX – Study Draft, dated 29 March 19.

Fortin, K. *The Accountability of Armed Groups under Human Rights Law*, Oxford: Oxford University Press, 2017.

Freier, N. *Outplayed: Regaining Strategic Initiative in the Gray Zone*, Carlisle, PA: United States Army War College Press, 2016.

Fremont-Barnes, G. (ed.). *A History of Counterinsurgency: From South Africa to Algeria, 1900-1954*, Santa Barbara: Praeger Security International, 2015.

Goldberg, A. E. *Bespredel* and the Conduct of Russian 'Hybrid Operations', *Small Wars Journal*, October 2016 https://smallwarsjournal.com/jrnl/art/bespredel-and-the-conduct-of-russian-'hybrid-operations'

Hansen, S. J. *Al-Shabaab in Somalia: The History and Ideology of a Militant Islamist Group, 2005-2012*, London: Hurst & Company, 2013.

Hicks, K., M. Matlaga and J. Schaus. *Zone Defense: Countering Competition in the Space between War and Peace*. Washington DC: Center for Strategic and International Studies, 2018.

HM Government, National Security Capability Review, March 2018. https://assets.publishing.service.gov.uk/government/uploads/system/uploads/attachment_data/file/705347/6.4391_CO_National-Security-Review_web.pdf.

House of Commons, The Report of the Iraq Enquiry: Executive Summary. Report to the Privy Counsellors'. 6 July 2016. https://assets.publishing.service.gov.uk/government/uploads/system/uploads/attachment_data/file/535421/The_Report_of_the_Iraq_Inquiry_-_Volume_VI.pdf

Idler, A. *Borderland Battles: Violence, Crime and Governance at the Edges of Colombia's War*, Oxford: Oxford University Press, 2019.

International Committee of the Red Cross. *The Geneva Conventions and their Commentaries.* https://www.icrc.org/en/war-and-law/treaties-customary-law/geneva-conventions.

International Committee of the Red Cross. *Montreux Document on Private Military and Security Companies.* https://www.icrc.org/en/publication/0996-montreux-document-private-military-and-security-companies.

International Court of Justice. *Military and paramilitary Activities in and against Nicaragua (Nicaragua and the United States of America)* UCJ Rep 1986, 14 ('The Nicaragua case').

Katzman, K. *The PLO and its Factions*, Congressional Research Service Report for Congress, The Library of Congress. 2002 https://www.globalsecurity.org/military/library/report/crs/11562.pdf.

Kau, M. Y. M. and J. K. Leung. (eds.). *The Writings of Mao Zedong: 1949-1976: Volume I, September 1949 – December 1955*, New York: M.E. Sharpe Inc., 1986.

Langer, A. and G. K. Brown (eds.). *Timing and Sequencing of Post-Conflict Reconstruction and Peacebuilding*, Oxford: Oxford University Press, 2016.

Lee, K. and A. Sullivan. 'People's Republic of the United Nations: China's Emerging Revisionism in International Organizations', *Center for a New American Security*, May 2019 https://www.cnas.org/publications/reports/peoples-republic-of-the-united-nations.

Lister, T., S. Sebastian and C. Ward. '*Putins Private Army*', CNN, August 2018, https://edition.cnn.com/interactive/2019/08/africa/putins-private-army-car-intl/.

Losh, J. 'Putin's Angels: the bikers battling for Russia in Ukraine', *The Guardian,* January 29, 2016.

Lovelace, D. C. Jr. *Hybrid Warfare and the Gray Zone Threat,* Terrorism: Commentary on Security Documents Vol. 141, Oxford: Oxford University Press, 2016.

Mazarr, M. *Mastering the Gray Zone: Understanding a Changing Era of Conflict,* Carlisle: United States Army War College Press. 2015

McGregor, A. 'Jihad and the Rifle Alone: 'Abdullah' Azzam and the Islamist Revolution', *Journal of Conflict Studies,* 23:2, 2006.

Merat, A. 'The Saudis couldn't do it without us', *The Guardian,* 18 June 2018. https://www.theguardian.com/world/2019/jun/18/the-saudis-couldnt-do-it-without-us-the-uks-true-role-in-yemens-deadly-war

Milanovic, M. *'Russian agents charged with downing of MH17; MH 17 cases in Strasbourg'*, June 2019. https://www.ejiltalk.org/russian-agents-charged-with-downing-of-mh17-mh17-cases-in-strasbourg/.

Murray, D. *Human Rights Obligations of Non-State Armed Groups,* Oxford: Hart Publishing, 2016.

Norton-Taylor, R. 'Britain is at the centre of global mercenary industry says charity', *The Guardian,* 3 February 2016. https://www.theguardian.com/business/2016/feb/03/britain-g4s-at-centre-of-global-mercenary-industry-says-charity.

Rabasa, A. et al. *The Lessons of Mumbai,* Santa Monica: RAND, 2009.

Reynolds, N. and J. Watling. 'Conference Report Royal United Services Institute for Defence and Security Studies Land Warfare Conference Securing Competitive Advantage', *RUSI,* June 2019. https://rusi.org/publication/conference-reports/land-warfare-conference-securing-competitive-advantage.

Robinson, L. et al. *Modern Political Warfare: Current Practices and Possible Responses,* Santa Monica: RAND, 2018.

Rühle, M. 'Deterring Hybrid Threats: the need for a more Rational Debate', *NDC Policy Brief,* July 2019. http://www.ndc.nato.int/news/news.php?icode=1335.

Secrieu, S. 'Russia in the Western Balkans – Tactical Wins, Strategic Setbacks', *EUISS Brief,* July 2019. https://www.iss.europa.eu/content/russia-western-balkans

Shevchenko, V. 'Little green men' or 'Russian Invaders', BBC news online, 11 March 2014, https://www.bbc.co.uk/news/world-europe-26532154.

Sukhankin, S. *'Russia's hired guns in Africa'*, European Council on Foreign Relations, November 2018. https://www.ecfr.eu/article/commentary_russias_hired_guns_in_africa.

Smith, R. *The Utility of Force: The Art of War in the Modern World,* New York: Alfred A. Knopf, 2005.

Stern, J. and J. M. Berger. *ISIS: The State of Terror,* London: William Collins, 2015.

Wehry, F. *Insecurity and Governance Challenges in Southern Libya,* Carnegie Endowment for International Peace Report (2017). https://carnegieendowment.org/2017/03/30/insecurity-and-governance-challenges-in-southern-libya-pub-68451.

The UK Government's approach to Stabilisation: A guide for policy makers and practitioners, UK Stabilisation Unit, 2019.

TRADOC Pamphlet 525-31, 'The US Army in Multi-Domain Operations 2028', *US Army Training and Doctrine Command*, November 2018.

Thompson, J. and Thranert, O. (eds). *Strategic Trends 2019: Key Developments in Global Affairs*, Zürich: Centre for Security Studies, 2019.

Troeder, E. G. A. *Whole-of-Government Approach to Gray Zone Warfare*, Carlisle: Strategic Studies Institute and U.S. Army War College Press, May 2019.

United Nations General Assembly. Session 55, Supplement No. 10 (A/56/10), http://www.un.org/law/ilc/.

US Army Mosul Study Group. 'What the Battle for Mosul teaches the Force', September 2017 https://www.armyupress.army.mil/Portals/7/Primer-on-Urban-Operation/Documents/Mosul-Public-Release1.pdf.

von Einsiedel, S. et al. *Civil War Trends and the Changing Nature of Armed Conflict*, United Nations University Centre for Policy Research 2017. https://collections.unu.edu/eserv/UNU:6156/Civil_war_trends_UPDATED.pdf.

Winkler, T. 'The Copenhagen Process on Detainees: A necessity', *Nordic Journal of International Law* 78, 2010.

CONCLUSION

Violent Non-State Actors, Silences, Binaries, Multiplication of Sovereignties?

Norma Rossi and Malte Riemann[1]

Introduction

The chapters contained in this volume argue that the analysis of Violent Non-State Actors (VNSAs) is crucial for developing improved and more nuanced conceptual, empirical, normative and policy relevant understandings of contemporary forms of conflict. In doing so, it asked multiple pertinent questions. How are VNSAs defined and categorised given that in contemporary conflicts dichotomous binaries are increasingly blurred? How is VNSA behaviour evaluated and how do normative commitments to state-centrism colour our understanding of such actors? What kinds of relationships exist within VNSAs and between VNSAs and states? How should VNSAs be engaged and if necessary countered?

This book has approached these questions without, of course, pretending to exhaust the range of possible answers. What it does, instead, is to point the reader towards a wide array of crucial trends and debates in relation to VNSAs in contemporary conflict. In doing so it provides a multi-faceted perspective on VNSAs that draws on mixed methodologies and different epistemological approaches, that speak to a growing scholarship

1 This chapter solely reflects the views of the authors and not the views of the Royal Military Academy Sandhurst, the UK Ministry of Defence or the British Army.

on the role and function of VNSAs in twenty-first century warfare.[2] In this concluding chapter, we highlight the book's contribution to such scholarship by providing a brief review of the chapters. Following on from this, we discuss three crucial themes that cut across the different contributions. These are the problem of state-centrism, the emergent possibility of multiple sovereignties and the practical challenges of countering VNSAs.

VNSAs in Contemporary Conflict: Themes, Approaches, Perspectives

As Donette Murray articulates in the introduction to this volume, the diverse chapters that compose this collection can be divided into three main groups. The first dealing with the conceptual challenge of understanding, defining, and theorising VNSAs; the second offers a range of methodological and empirical perspectives on studying the relationship between VNSAs and conflict; and the final part contains a multiple perspectives chapter that draws on experience of three practitioners with personal involvement in tackling the issues posed by VNSAs. Before discussing common themes, this following section summarises the key arguments of each contribution.

The opening chapter turns its attention to VNSAs in Africa. Here, Caroline Varin analyses terrorists, rebels and warlords and examines how these challenge one of the central features of state sovereignty: territorial control. Varin shows how these three distinct VNSAs convergence around their growing territorial ambitions and their long-term strategic approaches for engaging with local populations and providing some level of governance. Notwithstanding differences between these groups, Varin exposes how the typological distinctions between these actors shows itself to be increasingly blurred, highlighting the centrality of context specific empirical analysis to grasp how each of these actors operate in providing illicit governance. This analysis contributes to the expanding research agenda examining VNSAs as 'territorial contenders' of the state.[3] By using a wide range of empirical

2 See e.g. David Kilcullen, *The Dragons and the Snakes: How the Rest Learned to Fight the West* (Oxford: Oxford University Press, 2020); Sean McFate, *The New Rules of War: Victory in the Age of Durable Disorder* (New York: Harper Collins, 2019); Vladimir Rauta 'Towards a typology of non-state actors in 'hybrid warfare': proxy, auxiliary, surrogate and affiliated forces', *Cambridge Review of International Affairs* (2019).

3 Douglas Lemke, 'Do International Rules and Norms Apply to Nonstate Actors?', *International Studies Review* 21:2 (2019), 210–224; Douglas Lemke and Charles Crabtree, 'Territorial Contenders in World Politics', *Journal of Conflict Resolution* 64:2–3 (2020), 518–544.

examples, including ISIS, Boko Haram and AQIM, Varin shows that, while these converge in their *de facto* focus on territorial control, they still diverge in relation to their aims and the methods employed to achieve these aims.

In chapter 2, Helene Olsen continues this conceptual investigation by turning to an illustrious figure within the VNSA category; the mercenary. In her analysis, Olsen explores how 'distinct discursive practices' facilitated 'the conceptual understanding of mercenaries as illegitimate fighters'. Drawing on 'speech act theory' and based on a reading of discourses dating back to the Middle Ages, she shows how these actors have been most violently objected to and de-legitimised in situations 'where their presence on the battlefield has threatened prevalent conceptions of the ideal polity'. Key in this process of de-legitimation are moral objections towards these actors, which Olsen subdivides into deontological and consequentialist. The de-ontological aspect pertains to moral objections brought against the mercenary as an individual actor, focusing on the mercenary's foreignness and monetary motivation. The consequentialist objection, on the other hand, emphasises the possible consequences the hiring polity might face when employing these actors. A study of mercenaries, therefore, not only exposes the de-legitimation of these actors throughout time, as has been done by other scholars investigating mercenaries from a normative perspective,[4] but, more importantly, exposes how the very process of de-legitimation allows us to investigate how societies construct their version of the ideal polity.

In chapter 3, Helle Malmvig's chapter questions the dichotomy between states and VNSAs by showing how the supposed line dividing these realms are blurred and how non-state actors in conflict operate across this division to pursue their objectives. She does so by focusing on the Syrian *shabihas* – armed civilian gangs turned into irregular militia groups. In her analysis of these actors, Malmvig challenges the idea that states and VNSAs find themselves in a zero-sum competition over the monopoly of the legitimate use of violence. Instead, she shows how VNSAs in Syria operate within and outside of state structures, exercising criminal governance as well as partaking in localised territorial control. These groups, Malmvig shows, provide, amongst other activities, governance, collect taxes and

4 Sarah Percy, *Mercenaries: The History of a Norm in International Relations* (Oxford: Oxford University Press, 2006); Ulrich Petersohn, 'Reframing the anti-mercenary norm: Private military and security companies and mercenarism', *International Journal* 69:4 (2014), 475–493; Matteo C. M. Casiraghi, 'Weak, Politicized, Absent: The Anti-Mercenary Norm in Italy and the United Kingdom, 1805–2017', *Journal of Global Security Studies*.

adjudicate justice and therefore 'perform the state' themselves. In Syria, therefore 'political authority is bargained and negotiated between the state and a multiplicity of armed non-state actors', which compete with the state and simultaneously reproduce it. Taken together, these three opening chapters expose the complexities surrounding the study of VNSAs; their different typologies, and their multifaceted relations with states.

Following on from these conceptual debates, in chapter 4 Jack Watling analyses how VNSAs construct their own ethical and legal boundaries with regards to the use of force. In contributing to a growing debate on the legal personality of VNSAs,[5] he draws on his fieldwork experience and interviews with the PMF in Iraq. His ethnographic account analyses how the PMF formed its own ethical and legal boundaries for conduct on the battlefield and also investigates how this non-state actor's 'law of armed conflict' relates to International Humanitarian Law. Watling challenges the state-centric approach to IHL, arguing for the need to involve VNSAs in the shaping and updating of IHL, as well as for the need to relate promotion of compliance to IHL to local cultures and different normative codes. Through his rich empirical account, Watling argues that an enhanced local ownership of IHL must be combined with efforts by intervening states to adapt and explain their actions in ways that can be understood and accepted by the local population in which the military intervention takes place. In pointing out the unavoidable role of VNSAs in contemporary conflict, Watling suggests that the price of not including these actors in the shaping of IHL poses serious challenges to its application.

In chapter 5, James Fargher analyses how states use the fight against VNSAs as a pretext to pursue their own strategic and geopolitical interests. He does so by focusing on the involvement of Saudi Arabia and the United Arab Emirates (UAE) in the Yemen conflict. Though concentrating on a specific case, Fargher's contribution speaks to a broader phenomenon that can also be observed in other conflicts, such as Ukraine, Syria, and Libya.[6] Fargher shows how the efforts of Saudi Arabia and the UAE in Yemen reflect both states broader strategic and political ambitions in the region. These go

5 Themis Tzimas, 'Self-Defense by Non-State Actors in States of Fragmented Authority', *Journal of Conflict and Security Law* 24:2 (2019), 175–199; Heike Krieger, 'International Law and Governance by Armed Groups: Caught in the Legitimacy Trap?', *Journal of Intervention and Statebuilding* 12:4 (2018), 563–583.

6 Andreas Krieg and Jean-Marc Rickli, 'Surrogate warfare: the art of war in the 21st century?', *Defence Studies* 18:2 (2018), 113–130; Kristina Kausch (2017) 'State and Non-State Alliances in the Middle East', *The International Spectator*, 52:3, 36-47; Vladimir Rauta, 'Towards a typology of non-state actors' (2019).

far beyond the destiny of the Yemeni government and the security of its people and instead are geared much more towards regional rivalry with Iran. By deploying the 'War on Terror' narrative, actions against the Houthi rebellion have been 'lumped together' with the fight against ISIS and other VNSAs in the region. This, Fargher shows, constitutes the backbone of a legitimation strategy, leading to the de facto control of strategic regions in Yemen by the Saudi-led coalition. This has wide ranging effects on the region, as the Gulf States have been successful in establishing ties with African states, including Eritrea, Somalia, and Djibouti, to consolidate and extend their political influence around the Gulf of Aden.

Chapter 6 builds the counterpart to Fargher's contribution. While his analysis reveals how states use the fight against VNSAs as a pretext to pursue their strategic objectives, Marina Miron shows how powerful states use VNSAs to pursue their objectives. She demonstrates this via Russia's involvement in the Syria conflict. Specifically, Miron exposes how Russia attempted to use the YPG to its own advantage and also how Russia instrumentalises the Wagner group, a Russia-based PMC, to achieve its objectives in areas of strategic interest. Miron shows that, although it is difficult to control VNSAs, using these actors offers strategic advantages to powerful states, specifically in maintaining domestic support as these actors allow states to circumvent the so called 'body bag syndrome'. While this point has been made regarding the use of private contractors by Western states,[7] Miron shows how the same logic governs Russia's use of these actors. In this regard, Miron challenges the widespread assumption that only Western states are governed by 'post heroic' logics that demand casualty awareness.[8]

In chapter 7, Abigail Watson brings the recent turn towards remote warfare, by which Western states' combat threats at a distance without committing 'boots on the ground' or involving their own national assets,

7 Christopher Kinsey, *Corporate Soldiers and International Security*. (London: Routledge, 2006); Mateo Taussig-Rubbo, 'The Value of Valor: Money, Medals and Military Labor', *North Dakota Law Review* 88 (2012), 283–32. Bianca Baggiarini, 'Drone Warfare and the Limits of Sacrifice'. *Journal of International Political Theory*, 11:1 (2015), 128–144; Malte Riemann and Norma Rossi, 'Outsourcing Death, Sacrifice, and Remembrance: The Socio-political effects of Remote Warfare'. In *Remote Warfare: Interdisciplinary Perspectives*, E-International Relations, 2020.
8 Herfried Münkler. 'The wars of the 21st century', *IRRC* 85:849 (March 2003); Christopher Coker, *Waging War without Warriors? The Changing Culture of Military Conflict* (Boulder, CO: Lynne Rienner, 2002).

to the fore.[9] In her analysis, she examines the inherent risks involved in a specific aspect of such a strategy; approaches that work 'by, with and through' VNSAs. In exploring this issue, Watson analyses the Anti-Islamic State Coalition, where this approach has been a dominant feature. Her analysis of this coalition exposes the dangers of training and equipping non-state armed groups. Working with such actors has the potential for undermining and weakening state structures and carries the danger of challenging peace-making efforts by emboldening spoilers and by further fragmenting divided societies. Watson therefore ends her chapter with a warning: 'only by acknowledging and addressing the dangers of working with non-state groups can countries like the US, France and the UK ensure that engagements do not lead to more instability and conflict in the future'.

The final chapter, based on a roundtable held at the third *Sandhurst Trends in International Conflict* Symposium, constitutes a policy-focused engagement between three practitioners and their views on VNSAs. In this chapter, we aimed at collecting the voices of three practitioners tasked with understanding and tackling VNSAs. We brought them into conversation by posing two set questions: What are the key practical challenges posed by violent non-state actors in modern conflict? What current and future trends do you see developing in relation to violent non-state actors in modern conflict?

In discussing these, various themes examined throughout the preceding chapters re-emerged. Julien Bastrup-Birk from the Stabilisation Unit (SU) highlights the dynamic nature of the threat posed by VNSAs. Not only has their presence become a dominant feature of and challenge in contemporary conflicts, he argues, but their persistence in post-conflict reconstruction also poses a serious challenge to the authority of recovering states. He specifically highlights the emerging importance of the political economy of conflict as an area of interest for the SU. In this the SU echoes a growing emphasis in academia on this matter.[10]

---

9 Emily Knowles and Abigail Watson, 'Remote Warfare: Lessons leanrt from Contemporary Theatres', 27th June 2018, https://www.oxfordresearchgroup.org.uk/remote-warfare-lessons-learned-from-contemporary-theatres; Tom Watts and Rublick Biegon, 'Conceptualising Remote Warfare: The Past, Present, and Future', 22nd May 2019, https://www.oxfordresearchgroup.org.uk/conceptualising-remote-warfare-the-past-present-and-future.

10 Christine Cheng, 'Private and Public Interests: Informal Actors, Informal Influence, and Economic Order after War', *In Political Economy of Statebuilding: Power after Peace*, edited by M. Berdal and D. Zaum, 63–78. Abingdon: Routledge, 2013; Michael Dziedzic, *Criminalized Power Structures* (London: Rowman and Littlefield, 2016).

The military perspective offered by Matthew Walker and Grant Davies discusses VNSAs in relation to the 'grey zone'. 'Greyness' here indicates the presence of zones of legal and strategic ambiguity. Davies shows how VNSAs use the increasingly blurred lines between war and peace to their advantage, especially by exploiting zones of legal ambiguity. In these 'grey zone's, the making of (legal) decisions concerning who is to be held accountable, under what legal regime and by whom are key challenges. Davies explores these in relation to two aspects. First, how the line between legal and semi-legal PMCs are consciously kept vague by states employing the services of these actors; and secondly by calling attention to the challenges surrounding the legal parameters for handling captured persons within non-international armed conflicts. Walker's evaluation of the 'grey zone' instead focuses on how contemporary warfare is increasingly fought 'below the threshold of traditional conventional conflict'. Here, Walker shows how VNSAs provide adversarial states with tactical and operational agility, which can be translated into strategic gains, if not countered appropriately. Walker, therefore, argues that VNSAs are of key concern within the evolving context of (re)emerging great power competition.

Assumptions, Silences and Practices

By bringing the different contributions into conversation, we make three key points. First, we argue that these contributions challenge the assumption of state-centrism, that structures much of the literature on VNSAs,[11] at both the empirical and epistemological level. Second, in questioning state-centrism, we show how the growing role of VNSAs in contemporary conflict can be read as a symptomatic expression of processes leading to the multiplication of sovereign centres. By highlighting the multiplication of sovereignties, this volume contributes to attempts aimed at unravelling a key silent assumption within international relations, namely that sovereignty is an exclusive feature of states. Third, by reflecting on the policy implications of these claims, we aim at bridging the academic/practitioner divide and

11 Jens Bartelson, *Sovereignty as symbolic form* (London: Routledge, 2014).
Anna Leander, 'From Cookbooks to Encyclopaedia in the Making: Methodological Perspectives for Research of Non-state Actors and Processes', In *Researching Non-state Actors in International Security: Theory and Practice*, edited by Kruck, Andreas, and Andrea Schneiker (Abingdon: Routledge, 2017), 233–244.

through this highlight crucial conundrums states face in dealing with VNSAs in contemporary conflict.

State-Centrism

The debate about VNSAs in contemporary conflict has partially revolved around attempts to overcome silent assumptions at play within traditional accounts of International Relations. And here, state-centrism has been at the centre of the constitutive assumptions that VNSAs challenge.[12] State-centrism denotes the focus on the state as the sole, or at least key, actor in international politics. This state-centric notion becomes especially pertinent in relation to the conduct of conflict. While the legal framework has become increasingly nuanced and complex,[13] the political and normative dimensions of conflict remain tied to statist moorings. The chapters in this volume, explicitly or implicitly, address some of the problems that emanate from such a state-centric perspective in understanding contemporary conflict. In their approach to challenging state-centrism, the contributions can be divided into two groups. The first, including the chapters of Fargher, Miron, Varin, Watson and the three practitioners, do so at the empirical level. These contributions expose the plurality of actors involved in and their effects on contemporary conflict, highlighting the limitations of an exclusively state-centric perspective. Where these chapters challenge state-centrism at the empirical level, a second group questions the state and state-centrism at the epistemological level. Here, Malmvig, Olsen and Watling expose the contested process of drawing the line between state and non-state, as well as the politics involved in doing so.[14]

In the first group of chapters the state forms the analytical and normative point of departure. In relation to the regional politics of the Gulf States, and

12 Bartelson: *Sovereignty as Symbolic Form;* Malte Riemann "As Old as War Itself'? Historicizing the Universal Mercenary', *Journal of Global Security Studies,* 2020; Michael N. Schmitt, Sean Watts, 'Beyond State-Centrism: International Law and Non-state Actors in Cyberspace', *Journal of Conflict and Security Law* 21:3 (Winter 2016), 595–611; Matthew Norton, 'Real Mythic Histories: Circulatory Networks and State-Centrism', In *Global Historical Sociology,* edited by Go, Julian, and George Lawson (Cambridge: Cambridge University Press, 2017), 37–57.; Leander 'From Cookbooks to Encyclopaedia in the Making', 233–244. Tarak Barkawi, 2010. 'State and armed force in International context', In *Mercenaries, Pirates, Bandits and Empires,* edited by Colás, Alejandro and Bryan Mabee (New York: Columbia University Press, 2010), 33–54.

13 Władysław Czapliński, 'Recognition and International Legal Personality of Non-State Actors', *Pécs Journal of International and European Law* 1 (2016), 7–17.

14 Patricia Owens, 'Distinctions, distinctions: 'public' and 'private' force?', *International Affairs* 84:5 (2008), 977–990.

the Syrian conflict respectively, Fargher and Miron show how VNSAs act as an operational and tactical tool of state-centric power politics. By exposing the patron-client relations that structure the interactions between states and VNSAs, Fargher and Miron show how the non-state needs to be part of our analysis of state practices, even if these just form an extension of such. The three practitioners point towards the same issue. Walker warns how VNSAs are increasingly becoming a tool for states within a framework of re-emerging Great Power competition; Davies shows how VNSAs create zones of legal indistinction with the effect of blurring the state-centric lines between war and peace; and Bastrup-Birk highlights the importance of not losing sight of the challenges VNSAs pose to post-conflict reconstruction. Watson and Varin turn our attention to the ways in which we conceptualize VNSAs. Varin shows that maintaining a simple binary between VNSAs and the state is inadequate for capturing the complexity of VNSAs. In doing so she speaks to a broader trend in the literature that highlights the importance for drawing distinctions between different types of VNSAs.[15] Conceptualising VNSAs, however, is not solely an analytical endeavour, as Watson's chapter expresses. In analysing the increasing turn towards remote warfare, she shows how a better understanding of VNSAs can serve as a tool for critiquing state responses and conceptualizing VNSAs therefore acquires an important political function. Taken together, these contributions show how state-centrism oversimplifies the empirical realities on the ground and draw our attention to the complex nature of VNSAs.

On the epistemological level, Malmvig, Olsen and Watling question state essentialism by asking: How do we know the 'state'? What is part of it and what is not? How are the lines between inclusion and exclusion drawn? By posing such questions, the state loses its epistemological privilege and becomes a 'disputed being'[16] itself in need of investigation. This allows Malmvig, Olsen and Watling to focus their attention on how the line between state and non-state is drawn in practice, showing that the distinction between state and non-state is itself a political process that legitimises some actors while de-legitimising others. Malmvig does so by

15 For example, the volume by Natasha Ezrow classifies VNSAs by focussing on their ideology and motivation, strategy and tactics, structure and recruitment, funding methods, and their power on the state and society. Natasha Ezrow, *Global Politics and Violent Non-state Actors* (London: Sage, 2017), 6.

16 Costas M. Constantinou, *States of Political Discourse*. (London and New York: Routledge, 2004).

following Mountz, who argues that '[i]n state narratives, the dominant image is coherence in governance when, in reality, the state is a collection of contradictory, improvised and often disrupted responses to events'. Her analysis of the actors involved in the Syria conflict confirms this, as '[t]he boundary between VNSAs and the state is often so fluid that what constitute state and non-state forces is continuously contested and performed by the many armed actors themselves'. Watling identifies similar tendencies, showing how actors like the PMF 'straddle the divide between state and non-state actors'. While these contributions expose the difficulties of drawing boundaries between state and non-state actors in contemporary conflict, Olsen's contribution challenges state centrism from a radical, constructivist perspective. Specifically, she argues how our very condition of possibility for thinking about the political fabric of our societies – for example the ideal polity – is shaped through the exclusion of VNSAs and in opposition to these. The ways in which we conceptualise VNSAs, therefore, has less to say about these actors and more so about how we conceptualise and engage with the state, even if this engagement is not explicitly made and remains silent. Taken together, Malmvig, Olsen and Watling encourage us to reflect on an important point raised by Fred Halliday in the last stages of the Cold War: 'The argument is not about whether we are or are not 'state centric but what we mean by the state'.[17]

Multiplication of Sovereignties

The contributions to this volume also confront the assumption of sovereignty solely residing within the state. VNSAs question the changing character of political authority – for example, who can legitimately claim it, under what circumstances and where.[18] In this respect, the VNSAs analysed in this volume, including warlords, militias, insurgents, terrorists, mercenaries and PMSCs, are important for the challenge they pose to state authority expressed in Weberian terms (for example the monopoly of the use of legitimate violence).[19] This is an especially crucial aspect in war-making as a long tradition of historical sociology has shown. Charles Tilly, Stephen

17 Fred Halliday, 'State and Society in International Relations: A Second Agenda', *Millennium* 16:2 (1987), 217.
18 Bartelson, *Sovereignty as symbolic form*, 5.
19 Ezrow, *Global Politics and Violent Non-state Actors*.; Ersel, Aydinli, 'Assessing violent nonstate actorness in global politics: a framework for analysis', *Cambridge Review of International Affairs* 28:3 (2013), 424-444.; Klejda Mulaj, *Violent non-state actors in world politics* (London: Hurst & Company, 2010).

Krasner and Michael S. Drake, to name but a few, have shown how war-making has been a key feature in the historical constitution of centralised state authority.[20] While state sovereignty still shapes contemporary claims about legitimate political authority, the presence and growing importance of VNSAs in contemporary conflict calls such an assumption into question.

Varin's chapter points us precisely in this direction. Her analysis of a range of VNSAs in Africa shows how these share an increasing focus on territorial control and illicit governance. In territories in which such groups operate, the Weberian definition of the sovereign state as the holder of the 'monopoly over the legitimate use of violence' shows itself to be a normative ambition without corresponding to empirical reality. Instead, a multiplicity of overlapping and, at times, competing centres of illicit governance emerge. In line with a growing literature on criminalized power structures and nonstate governance,[21] Varin argues that the border between economically and politically motivated violence is blurring, which has the effect of leading to an overlap between governance and illicit non-state activities. Malmvig's analysis of the Syrian conflict amplifies this point. Her definition of Syria as a 'a hybrid order of shifting nodes of governance' is illustrative. Territorial sovereignty, conventionally attributed solely to the state,[22] appears instead as an intricated network of overlapping authorities. They thus challenge key assumptions concerning sovereign statehood such as a fixed territory, a centralized authority and a population in the form of a national people. Instead, in this network of territorial fragmentation, moving and fluctuating centers of authority emerge. Sovereign statehood appears less like a stable monopoly over the legitimate use of violence and more like a careful (re)negotiation of limited control over people. In this constantly shifting and changing frame, VNSAs might best be understood as constituting alternative forms of sovereignty. VNSAs might therefore be read as indicators for a fragmentation of state sovereignty where non-state actors perform multiple authority functions. In this, the contributions to this volume speak to a growing body of literature framing VNSAs variously as

20 Charles Tilly, 1985. 'War making and State making as Organized Crime', In *Bringing the State Back In*, edited by Evans, Peter B, Dietrich Rueschemeyer, and Theda Skocpol (Cambridge: Cambridge University Press, 1985), 169-191.Drake S. Michael, *Problematics of Military Power: Government, Discipline and the Subject of Power* (London and Portland: Frank Cass, 2002).

21 Norma Rossi, 'Breaking the nexus: conceptualising 'illicit sovereigns'', *Global Crime* 15:3–4 (2014), 299–319.

22 Stuart Elden, *The birth of Territory* (Chicago and London: The University of Chicago Press, 2013).

'petty',[23] 'illicit'[24] or 'informal'[25] sovereigns. Here the relationship between VNSAs and the state can be expressed along the lines of a competition/cooperation spectrum.[26] Some of these actors, as Varin shows, are directly opposed to state control and transgress the order expressed by the state in which they operate. In doing so, VNSAs and the state find themselves in a zero-sum competition for sovereignty. However, as illustrated by Malmvig and Watling, VNSAs also imitate state institutions and constitute forms of localized governance, which interact with and integrate into the state. Malmvig's analysis of the conflict in Syria especially exposes this process by showing how VNSAs in Syria have become part of the (in)formal governance structures of the state. Additionally, as Watling and Davies show, VNSAs interact not only with state law, but also engage with international law. In doing so, VNSAs operate sometimes in line with, at others contrary to the legal conduct of states. At the other end of the zero-sum competition framework we find situations, such as those highlighted by Fargher and Miron in relation to Russia and the Gulf States, where VNSAs act as an extension of state sovereignty. Similarly, Watson shows, in her analysis of remote forms of warfare, how VNSAs can also become an extension of the forces of an externally intervening state.

Taken together, the contributions to this volume offer a rich and varied illustration of the ambivalent relationship between VNSAs and state sovereignty, exposing the multiple ways in which VNSAs can 'stabilize, change, or overthrow existing power relations and legal orders'.[27] They furthermore highlight how VNSAs and their *modus operandi* cannot be grasped by a fixed model or placed inside a codifiable grid. Rather than adhering to fixed categories, such as state/non-state, political/economic, domestic/international, the intricated relationships exposed by the contributions call for an engagement with the relational character of the state/non-state binary. In doing so, the different chapters pay testimony to the importance of contextual and empirical investigations in mapping and understanding these complex interactions.

---

23 Miriam Wenner, 'Functions of sovereign violence: contesting and establishing order in Darjeeling, India', *Political Geography*, 2020.

24 Rossi, 'Breaking the nexus: Conceptualising 'Illicit Sovereigns'', 301.

25 Thomas Blom Hansen and Finn Stepputat, 'Sovereignty Revisited', *Annual Review of Anthropology* 35 (2006), 295-315.

26 Rauta, 'Towards a typology of non-state actors in 'hybrid warfare'.

27 Miriam Wenner, 'Functions of sovereign violence: contesting and establishing order in Darjeeling, India', 2.

VNSAS and Conflict in Practice

This section sets the academic contributions to this volume in dialogue with the concerns raised by the three practitioners. In doing so it is guided by the question: How can academic concerns speak to the policy-oriented concerns of practitioners dealing with VNSAs in conflict? Here, Varin raises an important issue that serves as the starting point for our considerations. She states that '[e]ffective counter-terrorist and counter-insurgency strategies will need to begin with a precise understanding of the nature, ambitions and methods of the enemy'. Understanding the adversary is indeed a key challenge emerging from the practitioners' contributions. Walker calls our attention to the importance of understanding VNSA narratives in order to disrupt the often disproportionate effects these actors have, given their seemingly limited resources. This is ever more pertinent for Walker, as he sees the role of VNSAs in future conflicts being *part of* and *not separate from* the return of inter-state rivalry and great power competition. This predicament, as both Fargher and Miron expose, has already become reality in Yemen and Syria, where aspects of Great Power competition play out. Addressing the issues posed by VNSAs therefore, Walker argues for a holistic approach, where the military component forms only one part of a wider strategic outlook employing diplomatic, political, cultural and economic tools. The contributions by Watling and Olsen reiterate this view, pointing towards the importance of the normative dimension. To understand VNSAs, it is crucial not to solely focus on material capabilities, but also on the ability and inability of these actors to mobilise legitimacy. This aspect is important, both for intervening states and the states in which VNSAs operate. Here, Olsen's chapter is illuminative as it shows how Western visions of the 'ideal polity' depend relationally on the ways in which VNSAs are framed. Watling furthermore shows how VNSAs themselves have a role in shaping such visions and normative frameworks.

To rethink legitimacy beyond the dominant statist paradigm, in turn, has profound practical implications. From a practical, policy-oriented perspective, the question of considering VNSAs as valuable and reliable allies is indeed of key importance. Watson's chapter seems to deny this possibility from the outset as VNSAs by definition challenge state sovereignty. VNSAs can therefore only be temporary allies that might bring operational and tactical gains, which have to be balanced with high strategic costs. The practitioners in this volume seem to agree with this, although arguing that the use of VNSAs by states is destined to continue.

Both Bastrup-Birk and Davies suggest that the main incentives for states to make use of VNSAs is the lack of accountability that surrounds these. Especially, in reference to the ambiguous line separating PMCs and mercenaries, Davies speaks about the 'buy-out' liability advantage that states gain. This concerns not only the legal aspect but also the moral and ethical responsibilities. Similarly, Bastrup-Birk uses the term 'proxies' to explain the link between states and VNSAs, arguing that state use of VNSAs accelerates in contexts where states aim at maintaining plausible deniability or a low profile. This also speaks to another aspect, the limited ability of external actors in intervening successfully in such dynamics. Walker points to a specific aspect of this issue in his discussion of 'grey zone' conflicts. Here, VNSAs are functional to the strategic objectives of adversary states' in contexts of sub-threshold warfare. These considerations emerge clearly from Miron's chapter and Russia's use of PMCs. In this, Miron's analysis reveals something unexpected. Debates on the advantages PMCs offer Western states often revolve around the supposedly casualty shyness of liberal democracies in 'wars of choice'. Yet, Miron's analysis reveals that Russian reliance on PMCs in Syria are influenced by the same concerns attributed to the so-called 'post heroic' societies in the West.[28] Miron's analysis thus calls into question the Western-centric character of such assumptions, pointing towards the need for further investigation.

The shared strategic necessity of Western and non-Western powers to work with and through these actors, poses also an additional normative problem. As Olsen reminds us, objections to the use of VNSAs are of a moral nature, shaping the ways in which visions of the ideal political community are imagined and actualized. This poses a variety of challenging normative questions. How do our visions of the ideal polity change if the moral objection to the use of VNSAs is lifted in the name of necessity? How can plausible deniability be morally justified and does its justification undermine the image and identity of liberal democracies as champions of open and transparent societies? From this normative perspective, the practitioners point us towards a dilemma. If external intervening states are guided by both normative and material concerns in their engagement with VNSAs, the question of appropriate distance from and proximity to VNSAs becomes important. Watson and Watling highlight this issue in relation to the choices made by both the UK and the US on this matter. Their contributions show how intervening forces risk being at once too

28 Coker, *Waging War without Warriors?*.

close and too distant from local VNSAs. Watson illustrates how the short-term calculations of working closely with local VNSAs can have negative long-term strategic effects in relation to Human Security and post-conflict stabilisation. Inversely, Watling exposes how the conscious choice of maintaining distance from an allied actor, in this case the Iraqi PFA, has meant sacrificing the intervening force's ability to influence these actors and control their behaviour. Watling and Watson thus expose the paradox that both approaches could yield negative humanitarian effects and undermine the legitimacy of the intervening states and, consequently, the stabilisation effort.

Fargher's chapter engages with this dilemma in relation to changes in the strategic context following 9/11. Fargher shows how Global War on Terror narratives and the principle of Human Security were instrumentalised to justify the Gulf States interventions in Yemen, which found both tacit and open support from important global players, such as the UK and the US. This exposes two aspects. First, challenging state-centrism should not involve forsaking the state as a crucial player in contemporary conflict, although being embedded within a complex network of multiple sovereignties. Second, the interactions between state and non-state do not take place within conditions of anarchy but are embedded within broader structural hierarchies that shape global politics.[29] Fargher's chapter is illustrative of both aspects, as it pays attention to the unequal relations between states occupying a privileged positions internationally, such as the US and the UK, states dominating regionally, as is the case with Saudi Arabia and the UAE, and states like Yemen, which on whose territories these complex power dynamics play out. This poses also a strategic dilemma: what to do when the interests of regional powers diverge from those of Western powers? The options of the latter could be limited in contexts where, as Watson shows, Western states are increasingly prone to rely on their proxies for local control in order to limit the costs of war. This issue is made even more pressing, as Bastrup-Birk shows, due to the highly globalised nature of these actors and their abilities to compete with states even with relatively limited resources.

In conclusion, by bringing the views of both practitioners and academics into the conversation, this volume illustrates how normative and practical concerns are interrelated yet doing so can leave us with

29 Janice Bially-Mattern and Ayşe Zarakol, 'Hierarchies in World Politics', *International Organization* 70:3 (2016), 623–654.

more questions than answers. That is, however, not to say that there are no identifiable lessons learned. What the chapters expose is the increasing awareness that political decisions concerning approaches of dealing with VNSAs in situations of armed conflict cannot afford to be unidimensional, considering only strategic advantages, normative commitments or economic concerns. Understanding, combatting and engaging VNSAs is thus in need of multi-dimensional approaches based on context specific understandings. Such an approach needs to start from an awareness that VNSAs need to be included in considerations aimed at ending hostilities, but actors need to be simultaneously cautious with regards to VNSAs potential for intensifying conflict. This volume is a contribution to navigate what is at stake and what are the risks involved such an endeavour.

References

Aydinli, E. 'Assessing violent nonstate actorness in global politics: a framework for analysis', *Cambridge Review of International Affairs* 28:3, 2013.

Baggiarini, B. 'Drone Warfare and the Limits of Sacrifice', *Journal of International Political Theory* 11:1, 2015.

Barkawi T. 'State and armed force in International context'. In *Mercenaries, Pirates, Bandits and Empires*, edited by Colás, A. and B. Mabee, 33-54. New York: Columbia University Press, 2010.

Bartelson J. *Sovereignty as symbolic form*. London: Routledge, 2014.

Berdal, M. "The 'New Wars' thesis revisited', in *The Changing Character of War*. edited by Strachan H. and S. Scheipers. Oxford: Oxford University Press, 2011.

Bially-Mattern. J. and A. Zarakol 'Hierarchies in World Politics', *International Organization* 70:3, 2016.

Boeke, S. 'Al Qaeda in the Islamic Maghreb: Terrorism, insurgency, or organized crime?', *Small Wars and Insurgencies* 25:5, 2016.

Briscoe, I. *Non-conventional armed violence and non-state actors: challenges for mediation and humanitarian action*. Norwegian Peacebuilding Resource Centre, May 2013.

Brown, D., D. Murray, M. Riemann, N. Rossi and M.A. Smith. *War Amongst the People: Critical Assessments*. Havant: Howgate, 2019.

Casiraghi M. 'Weak, Politicized, Absent: The Anti-Mercenary Norm in Italy and the United Kingdom, 1805–2017', *Journal of Global Security Studies*, 2020

Coker C. *Waging War without Warriors? The Changing Culture of Military Conflict*. (Boulder, CO: Lynne Rienner, 2002.

Czapliński W. 'Recognition and International Legal Personality of Non-State Actors', *Pécs Journal of International and European Law* 1, 2016.

Constantinou, C. M. *States of Political Discourse*. London and New York: Routledge, 2004.

Drake M. S. *Problematics of Military Power: Government, Discipline and the Subject of Power*. London and Portland: Frank Cass, 2002.

Dziedzic M. *Criminalized Power Structures*. London: Rowman and Littlefield, 2016.

Elden S. *The birth of Territory*. Chicago and London: The University of Chicago Press, 2013.

Ezrow N. *Global Politics and Violent Non-state Actors*. London: Sage, 2017.

Halliday F. 'State and Society in International Relations: A Second Agenda', *Millennium* 16:2, 1987/

Hansen T.B. and F. Stepputat 'Sovereignty Revisited', *Annual Review of Anthropology* 35, 2006.

Kausch, K. 'State and Non-State Alliances in the Middle East', *The International Spectator* 52:3, 2017

Kilcullen D. *The Dragons and the Snakes: How the Rest Learned to Fight the West*. Oxford: Oxford University Press, 2020.

Kinsey C. *Corporate Soldiers and International Security: The Rise of Private Military Companies*. London: Routledge, 2006.

Knowles. E. and A. Watson, 'Remote Warfare: Lessons learnt from Contemporary Theatres' 27th June 2018, https://www.oxfordresearchgroup.org.uk/remote-warfare-lessons-learned-from-contemporary-theatres

Krieg A. and J-M. Rickli. 'Surrogate warfare: the art of war in the 21st century?', *Defence Studies* 18:2, 2018.

Krieger H. 'International Law and Governance by Armed Groups: Caught in the Legitimacy Trap?', *Journal of Intervention and Statebuilding* 12:4, 2018.

Leander A. 'From Cookbooks to Encyclopaedia in the Making: Methodological Perspectives for Research of Non-state Actors and Processes'. In *Researching Non-state Actors in International Security: Theory and Practice*, edited by Kruck A. and A. Schneiker, 233–244. Abingdon: Routledge, 2017.

Lemke D. 'Do International Rules and Norms Apply to Nonstate Actors?' *International Studies Review* 21:2, 2019.

Lemke D. and C. Crabtree. 'Territorial Contenders in World Politics', *Journal of Conflict Resolution* 64:2–3, 2020.

McFate S. *The New Rules of War: Victory in the Age of Durable Disorder*. New York: Harper Collins, 2019.

Mulaj K. *Violent non-state actors in world politics*. London: Hurst & Company, 2010.

Münkler H. 'The wars of the 21st century', *IRRC*, 85:849, March 2003.

Norton M. 'Real Mythic Histories: Circulatory Networks and State-Centrism', In *Global Historical Sociology*, edited by Go J. and G. Lawson, 37-57. Cambridge: Cambridge University Press, 2017.

Owens P. 'Distinctions, distinctions: 'public' and 'private' force?', *International Affairs* 84:5, 2008.

Percy S. *Mercenaries: The History of a Norm in International Relations*. Oxford: Oxford University Press, 2006.

Petersohn U. 'Reframing the anti-mercenary norm: Private military and security companies and mercenarism', *International Journal* 69:4, 2014.

Rauta V. 'Towards a typology of non-state actors in 'hybrid warfare': proxy, auxiliary, surrogate and affiliated forces', *Cambridge Review of International Affairs*, 2019.

Riemann M. "As Old as War Itself'? Historicizing the Universal Mercenary', *Journal of Global Security Studies*, 2020.

Riemann M. and N. Rossi. 'Outsourcing Death, Sacrifice, and Remembrance: The Socio-political effects of Remote Warfare'. In McKay, A. (ed.). *Remote Warfare: Interdisciplinary Perspectives*, E-International Relations, 2020.

Rossi, N. 'Breaking the nexus: conceptualising "illicit sovereigns"', *Global Crime* 15:3–4, 2014.

Schmitt M. N. and S. Watts 'Beyond State-Centrism: International Law and Non-state Actors in Cyberspace', *Journal of Conflict and Security Law* 21:3, Winter 2016.

Taussig-Rubbo M. 'The Value of Valor: Money, Medals and Military Labor', *North Dakota Law Review* 88, 2012.

Tilly C. 'War making and State making as Organized Crime', In *Bringing the State Back In*, edited by Evans P. B., D. Rueschemeyer, and T. Skocpol, 169–191. Cambridge: Cambridge University Press, 1985.

Tzimas T. 'Self-Defense by Non-State Actors in States of Fragmented Authority', *Journal of Conflict and Security Law* 24:2, 2019.

Watts T. and B. Rublick 'Conceptualising Remote Warfare: The Past, Present, and Future', 22nd May 2019, https://www.oxfordresearchgroup.org.uk/conceptualising-remote-warfare-the-past-present-and-future.

Wenner M. 'Functions of sovereign violence: contesting and establishing order in Darjeeling, India', *Political Geography* 77, 2020.

INDEX

CPSIA information can be obtained
at www.ICGtesting.com
Printed in the USA
LVHW080347121021
700151LV00004B/147

9 781912 440207